Equine Behaviour in Practice

Understanding Horses and Treating Behaviour Problems

Debbie Busby

CRC Press
Taylor & Francis Group
Boca Raton London

CRC Press is an imprint of the
Taylor & Francis Group, an **informa** business

Designed cover images:

Top left: Enjoying a pleasure ride as a partnership. Photograph by Bryan Roberts

Top right: Sam Campling on Scarlett Rose at Somerford Park horse trials, 2019. Photograph by Matt Nuttall Photography

Bottom left: Aliyah Woodland of the NAC, target training her horse Bee. Photograph by Benny Payne (B. V. Productions)

Bottom right: Bonny Mealand with three Konik ponies. Photograph by Jean Sinclair

First edition published 2025
by CRC Press
2385 NW Executive Center Drive, Suite 320, Boca Raton FL 33431

and by CRC Press
4 Park Square, Milton Park, Abingdon, Oxon, OX14 4RN

CRC Press is an imprint of Taylor & Francis Group, LLC

ISBN: 978-1-032-94715-0 (hbk)
ISBN: 978-1-032-94714-3 (pbk)
ISBN: 978-1-003-58139-0 (ebk)

DOI: 10.1201/9781003581390

Typeset in Sabon
by Apex CoVantage, LLC

Equine Behaviour in Practice

Combining an in-depth exploration of equine behaviour with clinical practice from a new progressive perspective, ***Equine Behaviour in Practice: Understanding Horses and Treating Behaviour Problems*** focuses not just on the horse's behaviour but also on the behaviour of the human interacting with the horse. Acclaimed Clinical Equine Behaviourist Debbie Busby delivers evidence-based knowledge that will equip readers with the understanding and skills to become effective equine behaviour consultants.

Supported by considered discussion of how and why horses sometimes struggle to cope, the book comprehensively explains how to consult on and treat a range of equine behaviour problems. Rather than providing a formulaic "here's the problem; here's how to fix it", Busby encourages readers to develop a critical biopsychosocial evaluation based on an understanding of how horses evolved, their ethological and physiological adaptations, how they learn, and how their welfare and wellbeing can be assessed.

KEY FEATURES:

- Part 1 explains theories of equine evolution and the thin veneer of domestication
- Part 2 describes psychological and physiological theories of how horses learn, and how this can be applied in training and behaviour modification
- Part 3 illustrates this with reference to relevant legislation and widely used animal welfare models, as well as assessments of abnormal and stereotypic behaviour and practical aspects of measuring welfare
- Part 4 turns to the theory and practice of behaviour consulting, in three chapters divided into elements of behavioural assessment and treatment planning
- Part 5 examines the use of psychopharmaceuticals in behaviour therapy, discussing the roles of the behaviour consultant and the referring veterinary surgeon

Bridging the gap between science and practice, this book is a must-read for equine behaviour and veterinary students, equine behaviour consultants and practitioners, horse trainers, interested leisure riders and competition riders. It will empower equine professionals to combine behavioural and welfare knowledge with counselling and coaching skills to support horse owners in a mutual flourishing, enriching the lives of both species.

For Parkwood Rustler, with love, apologies and gratitude.

Oh! If people knew what a comfort to a horse a light hand is, and how it keeps a good mouth and a good temper, they would surely not chuck, and drag, and pull at the rein as they often do.

From Anna Sewell, *Black Beauty*

Contents

About the author

Debbie Busby graduated with a first class degree in psychology before going on to gain a distinction for her MSc in applied animal behaviour and welfare from Newcastle University. She grew up with horses and learned to ride aged 7 on Badger, one of the last retired pit ponies in England. Her first experience of using positive reinforcement to influence behaviour came when she had to ride Badger down a busy dual carriageway into Manchester to farrier Jack Priday's forge and was immediately sent to the next-door shop to buy a quarter of mint imperials. Somewhat bemusedly thinking they were for her, she soon learned who their actual recipients were: the ponies! Later Debbie looked after championship-winning harness racing horses and competed in eventing and dressage after studying for British Horse Society exams. After this she concentrated on teaching, especially coaching to improve performance, and developed a popular empathic and encouraging teaching style. At the same time she developed a fascination with the behaviour problems she saw in her clients' horses. Throughout this time, Debbie was partnered by her beloved chestnut thoroughbred cross Rusty, originally a dangerous challenge but emerging as a treasured companion, about whom you're sure to read in the pages of her books and articles.

Debbie Busby founded her successful international equine behaviour consultancy in 2006, and as a clinical equine behaviourist she works with vets and caregivers to resolve complex equine behaviour problems, as well as acts as expert witness in equine legal cases. A full member of the Association of Pet Behaviour Counsellors and a Certificated Clinical Animal Behaviourist, Debbie was only the third practising equine behaviourist in the UK to qualify to register at the highest level with the Animal Behaviour and Training Council (ABTC) as a clinical animal behaviourist specialising in horses. She is a graduate member of the British Psychological Society and an academic member of the International Society for Equitation Science. As a registered member of the British Association for Counselling and Psychotherapy she practices therapeutically within a Transactional Analysis framework, and she has drawn on this modality to develop a model promoting effective communication between horse owners and animal practitioners.

Debbie's intention as a behaviour consultant is to help horses and their owners to cope with the challenges of multispecies living: the domesticated horse is the same animal that ran wild over plains and steppe; domestication came late in this animal's history. Sometimes the mismatch between what we expect of our horses and what they need for physical and psychological health can be overwhelming, and behavioural problems often follow. Through her knowledge, experience and expertise Debbie helps owners understand and resolve the difficulties their animals are experiencing, achieving this in a way that is psychologically and ethologically sound.

Debbie writes books and articles and delivers lectures, presentations and workshops internationally on all aspects of equine behaviour, consulting and human behaviour change. She is a primary author of a book and a chapter contributor of another book. She has had articles published in a number of peer-reviewed journals and speaks at academic and professional conferences around the world.

As well as working with horses in the UK, Debbie travels to Spain, Turkey and the Middle East to research equine welfare and training practices and to consult with horse owners. Until the demands of her consultancy work took over, Debbie partnered with a trail riding company in Wadi Rum, Jordan, to organise week-long desert rides, sleeping under the stars each night and crossing the spectacular desert of mountains on strong, gentle, family-bred Arab horses.

Debbie supervises university students on behaviour courses at master's level and mentors new behaviourists coming through professional routes. She is currently conducting interdisciplinary doctoral research in horse-human relationships at Manchester Metropolitan University in the contexts of riding, sustainability and interspecies wellbeing.

Introduction

This book was birthed in the last century when I didn't know what to do with my new and very problematic young horse. The five-year-old I bought as a "proven showjumper" turned out to be nothing of the sort, and in fact hardly familiar with any aspects of the world that I'd bought him for. Bolting, rearing and flipping over, he was rapidly becoming a physical danger to both of us, and I realised I didn't know enough to be able to help him. I needed an equine behaviourist, but at that time, as a profession they didn't exist. Luckily, however, I received excellent advice from someone who understood the principles of supporting a horse to achieve a level of calm and relaxed wellbeing that would allow them to cope with their owner's requests and enable them to learn effectively. Trusty Rusty and I went from strength to strength in a range of disciplines from long distance rides through county showing to eventing and dressage, and when he left me at age 32 he owed me nothing, and I owed him 27 years of companionship, success and enjoyment.

Thus began the seed of an idea that eventually resulted in my professional certification, and a very fulfilling career, as a clinical equine behaviourist. And if you've picked up this book, you might be an aspiring practitioner wondering how you can achieve the same thing; where is the "how to be an equine behaviourist" handbook? Perhaps you're preparing for professional assessment, or you're studying welfare, veterinary or equine science and you've arrived at your behaviour-related assignments, or you're on a coaching or equine charity career pathway and you need to meet behaviour-related knowledge and understanding criteria. You want relevant information about everything you need to know to be an effective practitioner, but you haven't been able to find it all together one book. How do you get from the dawn of equine evolution, through ethology, biology and psychology (horse and human) to the nitty gritty of a one-to-one consultation with a distraught owner and a distressed horse? And how do you learn to develop your practice so that you meet all the knowledge, understanding and applied criteria set by the professional bodies that you hope to be registered with? It's my intention to meet a clear need within the equine behaviour sector to deliver soundly based knowledge that will equip students and current practitioners alike with the understanding to enable them to be effective and professional equine behaviour consultants.

Whether you're an aspiring behaviour practitioner, a student or a behaviour-curious owner or rider, in this single book you'll get an overview of who The Horse really is, their purpose as a species, their *telos* if you will, to draw on the Aristotelian perspective that "nature does nothing pointlessly", and what every horse needs to be to live a good life and be able to adapt their natural capacities to cope with life with humans. Everything we ask a captive managed horse to do depends on them being able to learn effectively, whether that's habituating to all the paraphernalia and experiences that form so much of their lives or how they learn to jump a Grand Prix track or perform advanced dressage. You also need to know how that happens, how you can work within best training practice and, crucially, what's to be done if things go awry and a horse shows, through changes in behaviour, that they aren't coping with what is being asked of them. Just as importantly, if you're a behaviour consultant, you need to achieve behavioural change in one species, the horse, through the vehicle of another species, their human guardian. Imagine how helpful it will be, therefore, to understand the human element of your work.

Numbers of publications provide current, evidence-based information about the behaviour of horses, but considered discussions of how and why horses sometimes struggle to cope with the demands placed on them by human guardians are less prevalent, and a comprehensive description and explanation of how to consult on and treat equine behaviour problems less common still. Understanding the applied aspects of equine behaviour is the gap that I seek to fill with this book, enabling students and behaviour practitioners to develop their knowledge and understanding of equine behaviour and, importantly, to apply that to the assessment and treatment of behaviour problems. Rather than providing a formulaic "here's the problem; here's how to fix it" (although I do offer a selection of worked examples), this book allows you to develop a critical biopsychosocial evaluation based on an understanding of how horses evolved, their ethological and physiological adaptations, how they learn, and how you can assess their welfare and wellbeing. The book then explains, in theory and in practice, how to apply this deeper understanding to behaviour problems so that as a clinical practitioner you can develop a safe, effective, ethical and legally compliant treatment plan and, crucially, you can pass your knowledge on to the caregiver in terms of behavioural awareness, application of recommendations and assessment of progress.

Part 1 begins with an explanation of theories of equine evolution and the thin veneer of domestication. I discuss the function and development of adaptive behaviour and the effects of how the horse's natural repertoire of behaviours interacts with the conditions of human guardianship and human-centric goals. Part 2 describes and explains psychological and physiological theories of how horses learn, and how this can be applied in training and behaviour modification. I explain how stress develops, the interactions between health and behaviour and the importance of considering disease, dysfunction and the effect of pain in behaviour modification. In the context of legally compliant and ethical practice an understanding of equine welfare, its standards and shortcomings, is essential, and Part 3 illustrates this with reference to relevant legislation and widely used animal welfare models, as well as assessments of abnormal and stereotypic behaviour and practical aspects of measuring welfare.

In Part 4 I turn to the theory and practice of behaviour consulting, in three chapters divided into elements of behavioural assessment and treatment planning. The following chapter in this section is dedicated to the human side of behaviour consulting, including models of counselling and communication, coaching and report writing, and finally I discuss what's important for non-vet practitioners to understand about behavioural psychopharmacology.

What if I hadn't received that help from an experienced friend in the first six months of my life with Rusty? What would have happened to him, a frightened, confused and, frankly, at the time, a dangerous young horse with 20 plus years ahead of him, his life completely determined by the decisions of people with unknown personalities and motivations? I would have missed out on 27 years of fun and fulfilment with him, and I may never have developed what became a lifelong interest in who horses are and why they do what they do. It's unusual for a horse to stay with one owner for that length of time, and as you'll read in this book, repeated moves of owners and yards is not one of the things that promotes good welfare. Most of my professional equine behaviour education took place after Rusty had left me, and not a day goes by when I wish that I knew when I first bought him what I know now. What a difference I could have made to his life. This is what I want for you if you are a behaviour student or an early career behaviourist; I want you to be able to help the Debbies and Rustys that enter your orbit, in a way that's science-based but that also allows you to critically consider not only the broad evidence base but also the relationality and individuality of the horse and their person in front of you, supported also by the unique understanding I also bring to the human element as a counsellor and psychotherapist. The decision to write this book comes as I reflect on a 40-year career working with horses in various capacities, and almost 20 years as an equine behaviourist, having helped many hundreds of one-to-one clients and many more by indirect routes. As you'll read in the communication and counselling chapter, clinical supervision is dear to my heart, and nowadays a lot of my work involves supervising and mentoring new and established practitioners, as well as assessing applicant clinical behaviourists for professional bodies. As a professional consultant, conference speaker and lecturer, I've worked in, and presented on, all aspects of equine and human behaviour change across five continents, and my book and article publications cover horse-human relationships (which is also my PhD research area) and my behavioural approach to equestrian disciplines. My most recent invitation to contribute to a government animal welfare network links me all the way back to a former career as a policy adviser and provides another pathway to influence national equine welfare decisions, an opportunity that could also be available to you as you progress in your equine behaviour career.

From Tinbergen to Transactional Analysis, in this unique text book for equine students and behaviour practitioners, whether aspiring or in practice, I aim to combine an in-depth exploration of equine behaviour and clinical practice from a new and progressive perspective, allowing you to gain your own deep and authentic understanding of how you can combine behavioural and welfare knowledge with counselling and coaching skills to support horse guardians/owners in a mutual flourishing that enriches the lives of both species.

Acknowledgements

It takes a team to write a book, and my immediate gratitude goes to Alice Oven for inviting me to write this book, and for her support and that of Shikha Garg throughout the writing and editing process. To my writing coach and picture researchers who got me here, I am indebted to you. My thanks also go to Valerie and Danielle Neff who first taught me to understand horses, to Dr Helen Spence who guided me in the right academic direction at a critical crossroads, and to Kelly Taylor-Saunders for your strength, understanding and humour through long years of behaviour debate and discussion. Thank you to Heather and Ross Simpson who developed the equine behaviour qualification that transformed my understanding of horses, their learning and their emotional lives. To Catherine Bell and Suzanne Rogers, thank you for your support and for research and authorship opportunities. I am grateful to Dr Emma Creighton, whose applied animal behaviour and welfare master's degree course at Newcastle University gave me the depth of knowledge and critical perspective I needed to push my behavioural understanding into new fields. To my colleagues in the Association of Pet Behaviour Counsellors and the Fellowship of Animal Behaviour Clinicians, your continual support and encouragement drives me on. Thank you to Carol Lucas for passing on your deep understanding of Transactional Analysis, allowing me to find new ways of reaching individuals and to help more horse-human partnerships to flourish in their good lives. My equestrian life has been made all the richer by the generosity of Bedouin horsemen Saleem and Salem Al Zalabieh who allowed me for so many years to share their beautiful Wadi Rum desert, and for that I am indebted. My horse and human clients, you have taught me more than I have probably passed on to you. Finally, thank you my dear Badger, Flair, Rusty, Ghalia and Kafkas; your tolerance, friendship and willingness on our many adventures gave me the deepest understandings of our interspecies relationality and what we inadequately describe as equine behaviour.

Part 1

Who is the horse?

Chapter 1

From evolution to domestication

The evolution of the horse

An appreciation of the horse's evolution is essential to our understanding of the modern horse. Why? Because the horse, like all living things, is a product of their evolutionary history and therefore that is the most significant factor in shaping their behaviour. It's also important to recognise that evolution is not about the creation of a "perfect" species, but about adaptation. The horse's survival depended on an ability to adapt in order to exploit a diverse range of habitats allowing successful opportunities to rear young and to avoid predation. Equally, evolution is not a linear process, with the current version improving on the last version. In reality each evolutionarily distinct individual existed because they were well adapted to their environment at the time. What this means is that, far from today's horse being a product of constant refinement, improvement and progression, the horse is actually a product of a changing and unpredictable climate. Those that went before were perfectly suited to the environment they lived in and flourished for millions of years until abrupt changes in climate caused their extinction.

The impact of diet

Much of the evolution of the modern horse – their size, speed and intelligence – can be explained by diet and changes in diet linked to these climatic changes. The earliest equids were browsers who ate leaves, shoots and woody plants, all of which are relatively high in nutritive value compared to grass and extremely abundant at this stage in earth's history. As the world continued to change so the horse adapted both physically and behaviourally, and that adaptability proved to be fundamental to the longevity of the species. Although today we have only one surviving genus (*Equus*), the fossil record shows that at their most successful there were at least 13 equid genera which included around 30 separate species. We'll go into more detail about taxonomic classifications later but suffice it to say that the predecessors of our modern horse were extremely good at adapting to and exploiting the environment in which they found themselves. Try to keep this in mind when we discuss the definition of normal and abnormal behaviour in later chapters. Some aspects of equine behaviour that are often labelled "abnormal" are the exact behaviours that allowed the horse to successfully adapt to new and changing environmental resources and thus ensured survival for the last 55 million years.

DOI: 10.1201/9781003581390-2

Rather than go into the finer detail of the many branches of the evolutionary tree (on which there are many good books, and which is illustrated at Figure 1.1) it is more helpful, for our purposes, to track the evolution of the horse through the progressive changes in anatomical characteristics over time, as this can give us an insight into how the horse's behaviour has developed. The generally recognised ancestor of the earliest equids existed in the tropical forests of the Eocene some 50–55 million years ago. Over the next five million years these animals adapted to the development of temperate woodlands, followed by a combination of woodland/savanna conditions until finally around 50 million years ago when seasonal, arid grasslands appeared. As these large grasslands took shape the horse developed traits that allowed them to exploit this habitat efficiently, so efficiently in fact that as the global climate became drier and forests were displaced by savannas and grasslands, there was a widespread burst of diversification in equids. The smaller browsing animals declined in number and branches of the equid species began to increase in body size from 25–50 kg to 75–100 kg and start grazing. The grazers required a more shearing action to grind food than browsing and so the depth and size of the horse's jaw increased to make room for more powerful muscles. Tooth durability was enhanced by a change from premolars to full molars, increasing the surface area available for grinding, and teeth became cement covered, higher crowned (hypsodont) and eventually ever growing (hypselodont). Grass is not only tough, but the cellulose that contains most of its available nutrients is locked away in the plant's cell walls which requires a lot of chewing to release. Food eaten off the ground is already full of dirt which means herbivores need to have much tougher teeth than browsers have, and not only did the teeth have to adapt but so too did the digestive system, which is what helped to drive the change in size.

Analysis of mitochondrial DNA (the DNA inherited from the mother) shows that for 30 million years the body size of equids remained relatively constant until around 20 million years ago when the species diverged into larger and smaller branches. The larger branches were the ones that survived to become the horse we know today (Goodwin, 2007). An important consideration in explaining why horses got bigger is the nature of the horse's diet and digestive system. Mammals are fundamentally incapable of breaking down the complex sugars (cellulose) found in the cell walls of plants. This problem is solved in the horse by forming a symbiotic relationship with the gut bacteria that break down the cellulose into a form the animals' enzymes can digest. It's a lengthy process that requires the animal to have a fermentation chamber where large amounts of plant matter can be stored while the bacteria do their job. Hindgut fermentation is the more familiar term we use today to describe this process. How does this link to size? Well, animal behaviour studies have shown that equids evolved to thrive on a diet consisting of poor or declining quality of forage by simply eating more. Hence the evolutionary pressure to increase size. Small animals need a lot more energy compared to their body weight than larger animals so in terms of efficiency it is better to be bigger if you need to eat greater quantities of food to compensate for a lack of quality (Budiansky, 2002).

During the Eocene the tiny forest-dwelling ancestor of today's horse would have been able to find a diet abundant with high-quality forage and very little fibrous cellulose. In the tropical climate of this epoch, young plant shoots would grow all year round, but as the climate cooled and growth became seasonal a fibrous diet would have been all that was available at certain times of the year, making fermentation and an increase in body size a necessity for the evolving horse. Another small but significant change that appeared early on in the evolution of horses, and not seen in other herbivores in the Eocene, was the development of a larger brain, in particular the area of the brain responsible for learning. This may have helped the first horses develop increased tactile sensitivity of the lips allowing them to become a generalist, particularly in terms of their diet. This ability to adapt from eating twigs, seeds and foliage to herbage and grasses helped ensure the horse's long-term survival (Busby and Rutland, 2019). We'll talk more about the horse's ability to learn and the implications for training in later chapters.

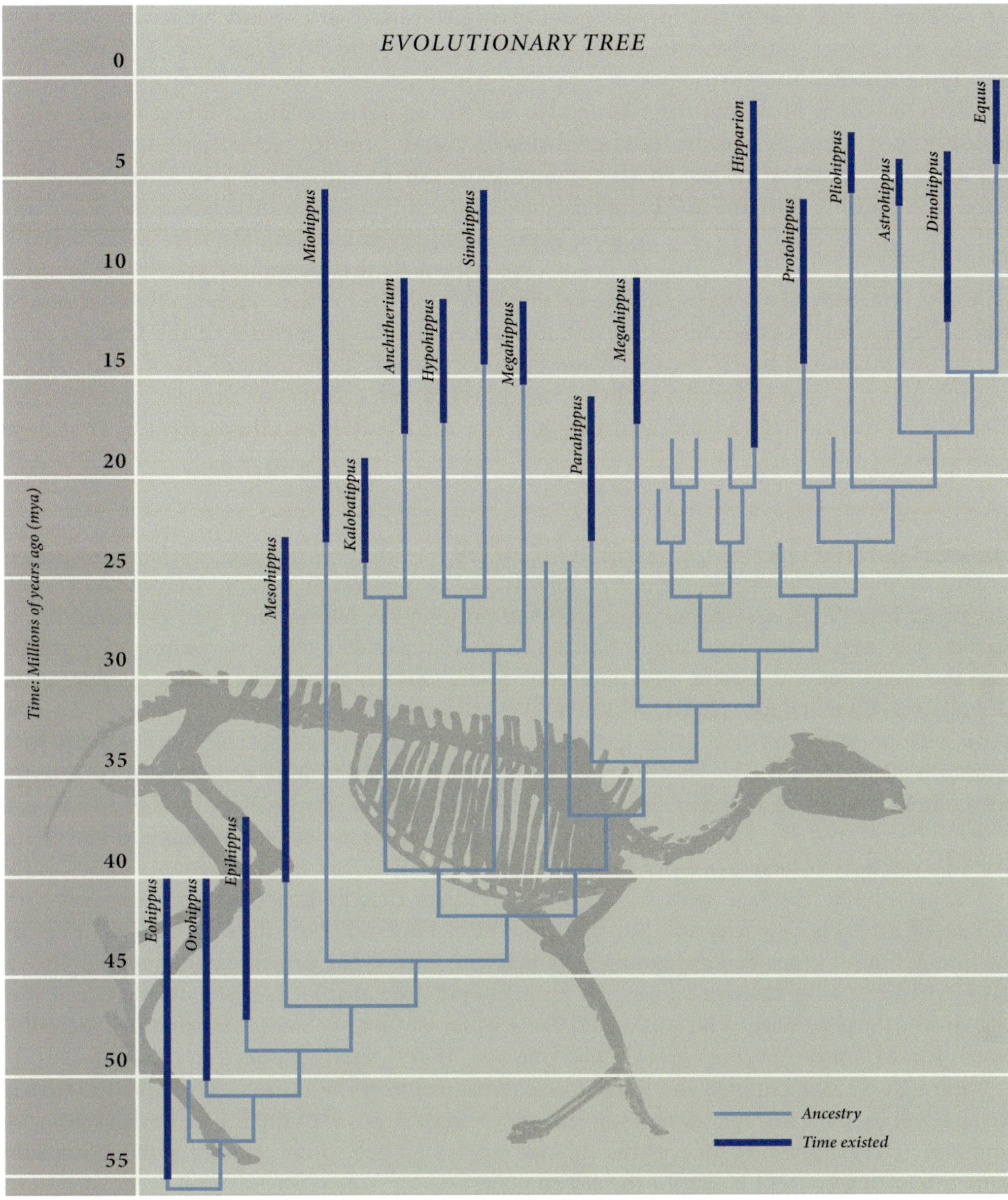

Figure 1.1 Evolutionary timeline of equid species, illustrating the divergence and development of different equids through various geological periods. Diagram by Nick Tcherniak.

The need to move quickly

There are disadvantages to being bigger. One consequence of increasing size is that getting up and down becomes harder work. A large herbivore in danger of predation is reluctant to lie down too much given the time it takes to get to its feet in an emergency. However, standing is also hard work (in a metabolic sense) for a large animal, hence the development of the stay apparatus which is unique to the horse and enables the locking of both the forelimbs and hindlimbs in place, removing the need to expend muscular energy in remaining upright. Interestingly, the horse can burn around 10% less energy standing up than lying down.

The move from browsing to grazing also led to the development of anatomical changes linked to movement, demonstrating how behaviour and form are inextricably linked. Longer legs allowed for longer strides, and the reduction of side toes made the lower limb lighter for more efficient limb movement. The appearance of the single-toed foot began around five million years ago in the Pliocene. The now plains dwelling, grazing equids were more exposed to predators and had to be able to flee. Development of the locomotor systems resulted in increased efficiency of the gaits and meant that although the horse is a relatively large animal, it is quite fast for its size and weight. At the same time the elongation of the face occurred which also facilitated timely escape. A long distance between the front of the mouth and the eyes allows an animal to simultaneously graze and keep an eye out for danger. The important lesson in all this is that size and swiftness are direct adaptations to the ecological challenges horses faced.

Development of social structures

The other interesting revelation from studying an animal's habitat and diet is the ability to predict their social structure. Small animals that are selective feeders in resource-rich forests tend to spread out and defend individual territories. This usually results in mating pairs being monogamous, a much more efficient system when prospective partners are spread over a potentially wide area. In contrast, large grassland animals that rely on patchier and scarcer food sources tend to form herds. This gives males the opportunity to hold harems and results in a polygamous system of reproduction. Equally, the move to a fibrous, low-quality forage diet which requires covering vast ranges in search of food made it inefficient for males to patrol and defend a territory for exclusive rights to a mate.

The take-home message with regard to evolution is that its purpose is not the long-term pursuit of progress or perfection but the short-term imperative of survival, and this requires the interaction of genes and environment (Goodwin, 2007). The pressures of climate upheaval (later combined with massive hunting by humans) put specialist feeders at a decided disadvantage. It was the generalists who could adapt to a wider variety of climate and circumstances that flourished. Migratory, curious species were the ones that thrived on upheaval.

Although we've examined the impact of environment on the horse's evolutionary journey, we have yet to make any mention of natural selection or survival of the fittest. Both of these are common in any discussion on evolution; however, from a behaviour point of view perhaps the more interesting concept is that of the selfish gene.

The selfish gene theory

The selfish gene (Dawkins, 2016) is an evolutionary theory proposed by ethologist Richard Dawkins to explain why animals choose to live together in groups over and above the need to reduce the chances of getting eaten. At its simplest level, the previously accepted wisdom was that animals on the periphery of the group were at most danger of predation, so each individual aimed to move towards the centre of the group and thus herds are formed. Dawkins argues that although this is an important consideration, it did not explain the whole picture. The selfish gene theory suggests that the more closely two individuals are genetically related, the more sense it makes, from a genetic perspective, to behave cooperatively. In essence, if animals live together in groups their genes must get more benefit out of the association than they put in. The selfish gene theory explains that it is not the survival of the group that matters or indeed the survival of the individual animal but the survival of the gene that the animal is carrying. Thus what drives behaviour is the imperative for the gene to survive and to be reproduced. The standard evolutionary understanding of "survival of the fittest" is not actually about survival in the sense of the individual but about fitness for survival and reproduction of the genes of that particular species or population. Taken from that perspective it is clear that there are advantages to living in a group that go beyond "safety in numbers".

Altruism

Paradoxically, the selfish gene theory actually requires that animals behave altruistically, as opposed to selfishly, in order for their genes to survive and be reproduced. Altruistic behaviour has no apparent benefit for the individual concerned. The most widely used example of this is in penguin colonies: penguins huddle together during extreme weather and alternate those at the centre where it's warm and the chance of survival is greater with those on the outside where it's coldest and they're more likely to perish. Why are the penguins on the outside of the group acting altruistically by risking their own survival rather than fighting to stay in the shelter of the group's centre? The answer is that there appears to be a collective imperative that ensures individuals keep rotating between the inside and the outside of the group in order to enhance the survival and reproduction of their genes as a whole. However, this seemingly altruistic behaviour of individuals within the group, which has no benefit to them and potentially puts their own survival at risk, is also a selfish act when considered at the gene level. If the penguins never showed any altruism and did only what was best for them as individuals then it may be more likely that they would die out, first as an individual and then as a population.

Horses within a herd also demonstrate altruistic behaviour by, for example, standing tall and snorting (and thus making themselves noticeable to a predator) to alert the rest of the group when a predator is nearby. This seems counterintuitive in that the best thing for the individual, once they've noticed the threat, is to run away and take advantage of the fact that they saw it before the others. But that's not what happens. At the risk of drawing attention to themselves, one horse will raise the alarm to give the whole herd a chance to escape, even if it increases their risk of being eaten. Consider the alternative and perhaps this is truly where the influence of the selfish gene theory comes in. Suppose that a vigilant horse spots the predator lurking in the grass before all the others and chooses to simply run away without alerting the rest of the group. Now that horse is on their own and no longer benefits from the "safety in numbers" that a herd provides. Most predators will go for the single or lone animal rather than diving into the whole group. The horse's best chance of survival is to alert the group and become one of the many animals in the departing herd. If all members of the herd act in the same altruistic way they give themselves, the population and their gene pool the best chance of survival over the long term.

Reciprocal altruism

Reciprocal altruism, otherwise known as "you scratch my back and I'll scratch yours", is another facet of the selfish gene theory that is worth exploring in relation to understanding horse behaviour. Suppose a species of animal is parasitised by a particularly nasty tick which carries a disease. It's important that these ticks are removed as soon as possible. Normally an individual animal can pull off its own ticks whilst grooming itself. However, there are places that the animal is unable to reach unaided (think top of the withers for a horse). The solution to this problem is to recruit a friend to do it for you and later, when the friend is parasitised themselves, the good deed can be paid back. Hence, you scratch my back and I'll scratch yours. In fact, mutual grooming is a very common activity in both mammals and birds. What's interesting is that if you think about what is required for this simple good deed to become such a common activity you realise that there is often a delay between the good deed and the repayment. Which means, as the theory goes, that reciprocal altruism can only evolve in a species that is capable of recognising and remembering each other as individuals.

This recognition and recall is important. Suppose horse *B* has a tick on their withers and horse *A* pulls it off. Later horse *A* has a tick on their withers and will naturally seek out horse *B* to return the tick-pulling favour. However, *B* simply turns their nose up and walks off. According to Dawkins, *B* is a "cheat", an individual who accepts the benefit of another's altruism but who does not pay it back, and in a group cheats will generally do better than altruists because they gain the benefits without paying the cost. Consider if the group only consisted of "suckers", who groom anybody who needs it, and cheats, who accept altruism from the suckers but never groom anybody else. The group wouldn't last very long. The suckers would gradually die out as they succumbed to the disease the ticks they couldn't reach were carrying, whilst the cheats would follow soon after because there was nobody left to groom them. However, suppose there was a third set of individuals, what Dawkins called the "grudgers". Grudgers groom strangers and individuals who have previously groomed them. However, if any individual cheats them, they remember the incident, bear a grudge and refuse to groom that individual in the future. In this group scenario consisting of three sets of individuals – suckers, cheats and grudgers – the most successful strategy is that of the grudgers, and these are the individuals that are likely to thrive and survive. Since horses engage in mutual grooming and if we accept the theory of reciprocal altruism then we must consider that horses as a species are able to recognise and remember individuals as well as to distinguish between the cheats and the suckers. What might the implications of this be for us humans when we are working with and training our horses?

Evolutionary stable strategies

Because so many animals have evolved to live in groups, clearly benefit from being in groups and have continued to stay in groups over millions of years, we can say that this is an evolutionary stable strategy (ESS), a pattern of behaviour that is consistent and hasn't changed throughout the evolutionary journey. And the reason it hasn't changed is because it is a strategy that if most members of a population adopt it, it cannot be bettered by an alternative strategy. In other words, the best strategy for an individual depends on what the majority of the population is doing. Since the population consists of individuals each aiming to maximise their own success, the only strategy that persists is one which, once evolved, cannot be bettered by any outlier individual. Following a major environmental change there may be a brief period of evolutionary instability, but once an ESS is achieved it will remain, and selection will penalise any deviation from it. For horses, mutual grooming and living in a herd are evolutionary stable strategies. Until humans came along! Which leads us nicely into a discussion about domestication.

The impact of humans and domestication

Horses have been associated with humans for a relatively short period of time, particularly when you consider the millions of years over which the horse has evolved. So how did the horse's journey into domestication start and what does it mean to be domesticated?

Domestication is the process by which a population of animals becomes adapted to humans and to the captive environment. A tame animal is generally considered different from a wild one in that it is dependent on humans and will stay close to them of its own free will (Clutton-Brock, 1987). Archaeological and animal behaviour studies have situated domestication as a long slow process of mutual adaptation in which those animals that "hung" around human settlement gained more than they lost, including abundantly grown crops and protection from predators due to the proximity of humans. Those semi-domesticated horses that were more curious, less territorial, less aggressive, more dependent and better able to deflect human aggression through submission were able to flourish in this new environment. In a more recent scientific review Ahmad *et al.* (2020) posit three pathways to domestication: commensal, prey and direct. The commensal pathway describes the process outlined previously where wild animals are attracted to human settlements for the benefits of food, shelter or protection, thus establishing a commensal relationship with people. Dogs, cats and chickens are generally cited as following this pathway to domestication. In the prey pathway, humans start hunting species like pigs and cattle for their meat. Over time the development of game management strategies to ensure the local stock of these animals were not depleted morphed into controlled breeding according to traits desirable to humans. The direct pathway is characterised by the capture of wild animals (horses, donkeys and camelids) for meat, milk, wool and transportation and controlling their movements, nutrition and reproduction.

Researchers continue to explore what it means to be domesticated (Puruggan, 2022). It is easier to interpret physical evidence such as bones found in archaeological digs and paintings on walls, as in Figure 1.2 than it is to speculate about a time in history with no written records or physical artefacts and extrapolate that into a sequence of events that may have led to domestication. It is also quite common to talk about domestication in relation to human intentions and intervention, which minimises the role of unconscious selection. The broader definition of domestication describes it as a coevolution process that arises from mutual benefit, situating domestication as an explicitly evolutionary process that arises from an ecological interaction, in essence a process where two or more species interact and evolve reciprocally through

Figure 1.2 Illustration of a horse in rock art, highlighting early human interaction with animal motifs. Photograph by Maxim Petrichuk.

selection and as a result both species derive a benefit. From a behavioural point of view it is useful to consider differing definitions and theories as there is valuable insight to be had from their differing perspectives.

Interestingly, the archaeological record shows that horses were domesticated thousands of years after the dog, sheep, goat and cow (Librado *et al.*, 2016) and this may reflect the fact that it would require considerable skill on the part of Neolithic humans to successfully domesticate a large, agile animal with a tendency to migrate over large distances. Despite this, the horse is still one of fewer than a dozen mammals, of the more than 4,000 species that have occupied the earth during the last 10,000 years, to have achieved widespread success as domesticated animals. Why is that? Perhaps, as Stephen Budiansky suggests in his book *The Nature of Horses* (2002), the horse is "preadapted" for a domestic life. Perhaps domestication had less to do with the intention or needs of humans and more to do with the quirks of adaptation and evolution that turned out to be vital to the horse's future ability to thrive and survive. The animals best able to exploit domestication were generalists – able to eat a variety of widely available foods and relatively non-territorial social animals not given to disruptive intraspecies conflict over fixed pieces of ground. Even before humans arrived the horse may have been well placed to make the best of what domestication would have to offer.

Humans began taking horses from environments in which they had evolved, and managing them under conditions which were convenient to humans, around 5,000 years ago, as evidenced by the abundant distribution of chopped bones found in the kitchen debris of early human settlements. These horses were initially maintained as a food source (meat and milk) in herded groups within their natural environment. Orphaned foals may well have been captured and kept as pets or traded as a food animal before supplanting the slower ox as a draught animal. The earliest direct evidence for true horse domestication that can be dated is found in the artistic, textual and funerary evidence towards the end of 3rd millennium BCE, whilst the first drawing of a horse being ridden appears around 1560 BCE. This coincides with physical evidence of horses being ridden found in graves alongside other artefacts dating to the beginning of the 2nd millennium BCE.

Over time, human activities have either directly or indirectly influenced the evolution of the horse and caused drastic reductions of truly wild populations. The Tarpan horse went extinct in 1909 and even Przewalski's horse (Figure 1.3) horse was overhunted to the extent that it was declared extinct in the wild with only a handful surviving in captivity. Happening in parallel with the extinction of wild populations, human-driven management through selection has significantly influenced the recent history of domestic horses through the development of multiple breeds. It was once people started breeding animals in captivity that human intervention had a more decisive and impactful role on their evolution. During domestication we were able to consciously select and place emphasis on traits that worked best for us and our needs – strength, aesthetics, speed and endurance. With the exceptions of the Arabian, Mongolian and Icelandic horses, breeds are a relatively recent human construct when looked at on an evolutionary timescale. According to the Food and Agricultural Organisation of the United Nations, of those multiple breeds developed by humans, 87 are already extinct, and among the remaining 905 almost a quarter are categorised as at risk (Librado *et al.*, 2016).

The differing changes we see between wild and domestic animals is likely the result of both genetic and developmental factors. Much of the research to date has been aimed at characterising global patterns of varying horse populations and identifying disease and its treatment. Genetic research has allowed scientists to explore how the modern horse differs from ancient ancestors, with genes linked to coat colour being the most widely studied, not least because genes linked to coat colour are the ones most associated with early domestication. It is only relatively recently that science has investigated further what it means to be a horse, both in terms of evolution and genetics and in terms of behaviour, perception and learning. Future studies are expected to increasingly focus on how the many past cultures and horse husbandry practices have impacted

Figure 1.3 Horse and human across the millennia: Bonny Mealand with one of a group of Takhi at Scotland's Highland Wildlife Park. Now often referred to by their Mongolian name, Takhi, or "spirit horse", it's clear that they bear significant differences in comparison to *Equus ferus caballus*. Photograph by Jean Sinclair.

on horse genetics, and there is already some suggestion that domestication results in the persistence of juvenile features within the individual at both the physical and behavioural level. This may be because the behaviour of young animals is more flexible and therefore better able to adapt to a life of domesticity, leading breeders to actively select for this trait.

The consequences of domestication

Domestication has required horses to live in close proximity to humans and to fit in with our lifestyle. Any animals that don't fit in tend to be culled or no longer chosen for breeding purposes and so the population changes, not as a result of natural selection or environmental changes but as a result of being more suited to this way of life. The pressures of natural selection and predation are reduced because humans care for the domesticated animal. Even the less fit horse and other animals that would not normally survive in the wild thrive and are looked after. However, when animals are kept in small spaces there is an increased risk of disease, and because they can't get away from each other there may be more aggression. Large domesticated groups also mean there is more opportunity for breeding, and this can intensify sexual behaviour. These closed, captive population groups increase the tendency for inbreeding and the appearance of traits normally masked by more dominant genes and therefore not seen in wild populations.

Today's horses tend to have their freedom to roam and choose food, shelter, mates and social companions restricted by human choices and decisions. We also expect them to accept and interact with us, and to understand our instruction even though we evolved as predators and our own behaviour has been shaped in a very different evolutionary journey. We ask horses to do many things which build on their instincts, especially their well-developed social instincts and their natural ability to learn, but we also expect them to do many things that are counterintuitive, such as walking into a dark trailer, carrying the weight of a rider on their backs and leaving their social group to go somewhere alone. The domestic horse is now wholly dependent on us for its survival, wellbeing and future. Whether one views this as a good or a bad thing probably depends on one's perception of "the equine condition" in the 21st century. We discussed the impact of abrupt climate change on the evolution of the horse earlier in the chapter, and it was one such abrupt change in climatic conditions 15,000 years ago (probably exacerbated by overhunting by humans) that almost drove the modern horse to extinction. Were it not for domestication, *Equus ferus caballus* would likely not be around today. This seems an appropriate time to move on and consider the current evolutionary stage of the horse and its relations.

Taxonomic classification

Taxonomy is the science of naming, describing and classifying organisms, and phylogeny studies their evolutionary history and relationships among individuals or groups of organisms. The combination of the two into taxonomic classification allows for the development of a hierarchical classification based on the sharing of physical and behavioural characteristics (Table 1.1). Members of the lower levels have more characteristics or traits in common than those on the upper levels. Hence the taxonomic classification of the modern domestic horse looks like this:

The Equidae family includes horses, zebras, asses and donkeys as well as all the now extinct ancestors of the horse. The genus *Equus* to which all living equids belong evolved a few million years ago and contain seven species including the modern horse. You'll notice from Table 1.2 that whilst horses, donkeys, asses and zebras fall under the same family (Equidae) and the same genus (*Equus*), they are distinct species.

Domesticated and wild horses are now considered a single species with the valid scientific name for the horse species being *Equus ferus*. You may have seen the domestic horse being referred to as *Equus caballus* in other texts; the reason for this is that in the past taxonomists considered domesticated horses and wild horses subspecies of the same species. Thus domesticated horses were labelled *Equus caballus caballus* and wild horses like the Przewalski's horse were labelled *Equus caballus przewalskii* (more shortly on whether Przewalski's horse is truly a wild horse). However, in 2003 the International Commission on Zoological Nomenclature decided that the scientific names of wild species have priority over the scientific names of domesticated species, hence mandating the use of *Equus ferus* for all species of the horse. *Equus ferus caballus* denotes the modern domestic horse with *Equus ferus przewalskii* denoting Przewalski's horse.

Table 1.1 Taxonomic Classification of the Modern Domestic Horse

Level of classification	Group that the horse belongs to	Characteristics
Kingdom	Animals	Share certain cellular features not found in plants, fungi or micro-organisms.
Phylum	Chordates	All chordates are animals with at least a primitive basis of a backbone at some stage in their development.
Class	Mammals	All mammals produce milk for their young, tend to grow hair and develop their young inside the body rather than in eggs.
Order	Perissodactyl	Mammals with either one or three hoofed toes on each hind foot; they are also hind gut fermenters.
Family	Equidae	Includes horses, zebras, asses and donkeys. All the ancestors of the horse also fall under this category.
Genus	Equus	The only group in the family Equidae that has survived to the present day.
Species	Ferus	Brings together all the various breeds of domestic and wild horse. Members of the same species are capable of breeding and produce fertile offspring.

Table 1.2 Equus Species and Subspecies

Genus	Species	Subspecies
Equus	Ferus	Domesticated horse (*Equus ferus caballus*) Tarpan (*Equus ferus ferus)* Przewalski's horse (*Equus ferus przewalskii*)
	Africanus	Nubian wild ass (*Equus africanus africanus*) Somalian wild ass (*Equus africanus somaliensis*) Atlas or Algerian wild ass (*Equus africanus atlanticus*) Modern donkey or ass (*Equus africanus asinus*)
	Hemionus	Syrian wild ass (*Equus hemionus hemippus*) Mongolian wild ass (*Equus hemionus hemionus*) Turkmenian wild ass (*Equus hemionus kulan*) Persian wild ass (*Equus hemionus onager*) Indian wild ass (*Equus hemionus khur*)
	Grevyi	Grevy's zebra (*Equus grevyi*), largest living wild equid. Weighs up to 450 kg and stands at 5 feet 2 inches high. Found in Kenya and presently listed as endangered.
	Quagga	Common plains zebra (*Equus quagga*), despite its name was classified as "near threatened" by the International Union for Conservation of Nature (IUCN) in 2016. Not to be confused with Quagga (*Equus quagga quagga*), hunted to extinction in the wild by the late 19th century.
	Kiang	Western kiang (*Equus kiang kiang*) Eastern kiang (*Equus kiang holderei*) Southern kiang (*Equus kiang polydon*)
	Zebra	Cape mountain zebra (*Equus zebra zebra*) Hartman's zebra (*Equus zebra hartmannae*), found in South Africa and is considered vulnerable.

In chapter 2 we move on to look at ethology, the scientific study of animal behaviour. Ethology allows researchers to develop an ethogram or catalogue of the different kinds of behaviour observed in an animal. I mention this here because it is important to bear in mind that whilst horses, donkeys and asses share the same taxonomic family, Equidae, because of their different routes of evolution they will each have separate and distinct catalogues of behaviour.

The African asses

The continent of Africa has produced only one domesticated mammal – the donkey. It is likely that by the 4th millennium BCE the ass was bred in captivity by the ancient Egyptians; certainly there are depictions of the domestic ass in wall paintings found in tombs dating to around 2500 BCE. Within the species *Equus africanus* there are four subspecies of African wild ass:

- Nubian wild ass (*Equus africanus africanus*) which is believed to be extinct.
- Somalian wild ass (*Equus africanus somaliensis*) which is near to extinction with only around 700 alive in the wild.
- Atlas or Algerian wild ass (*Equus africanus atlanticus*) which is now extinct.
- *Equus africanus asinus* – the modern donkey or ass (terms which can be used interchangeably) that was domesticated around 3,000 years ago and used both as a companion and working animal.

The Asiatic asses

The Asiatic wild ass or *Equus hemionus*, meaning "half donkey", is an ancient equid species that has existed for more than four million years. There are four remaining subspecies; the Syrian wild ass (*Equus hemionus hemippus*) having gone extinct in 1927:

- Mongolian wild ass (*Equus hemionus hemionus*)
- Turkmenian wild ass (*Equus hemionus kulan*)
- Persian wild ass (*Equus hemionus onager*)
- Indian wild ass (*Equus hemionus khur*)

Interestingly *Equus kiang*, another member of the Equidae family descended from this group, is the biggest wild ass and is found in parts of China, Nepal, India and Pakistan. Growing to around 4 feet 7 inches and living on lowlands, they predominantly eat grass but also roots and shrubs. They are considered a species of least concern probably due to the lack of predators.

The wild horse

Equus ferus (wild horse) contains three subspecies – the domesticated horse (*Equus ferus caballus*) which is the most common subspecies, the extinct Tarpan (*Equus ferus ferus*) and the endangered Przewalski's horse (*Equus ferus przewalskii*).

A genetic study by Gaunitz *et al.* in 2018 claimed that no truly wild horses still exist. When scientists talk about a species being truly wild, what they mean is that there is no history of domestication. In the case of the Przewalski's horses inhabiting the Mongolian grasslands, it is currently thought that they are a feral descendent of the earliest domesticated horse. It's estimated that Przewalski's and domestic horse populations diverged some 38–72 thousand years ago, and findings from the first genome sequencing research suggest that both populations likely remained isolated after that divergence. Genetic analysis demonstrated that Przewalski's horse is descended from the earliest known instance of horse domestication by the Botai people of northern Kazakhstan around 5,500 years ago. These Botai horses were not direct ancestors of modern domesticated breeds, but based on DNA findings they did give rise to Przewalski's horse. It is currently thought that horses from the domestic Botai herds escaped and became the Przewalski's horse "discovered" in the 19th century (Gaunitz *et al.*, 2018). Interestingly, this provides a good example of the prey-domestication pathway we discussed earlier in which the domestication incentive for settled (non-nomadic) Botai hunter-gatherers was to maintain and manage access to an animal that represented the core of their subsistence way of life.

Today, the current population of Przewalski's horse numbers around 2,000, all of which are descended from the 15 individuals captured a century ago and who were subsequently reintroduced into the wild. Recent research reveals that significant changes were introduced to the gene pool of those 15 animals whilst in captivity, hence they may no longer be considered truly wild (Orlando, 2020).

References

Ahmad, H. I., Ahmad, M. J., Jabbir, F., Ahmar, S., Ahmad, N., Elokil, A. A., & Chen, J. (2020). The domestication makeup: Evolution, survival, and challenges. *Frontiers in Ecology and Evolution*, *8*, 103.

Budiansky, S. (2002). *The Nature of Horses*. Guernsey: The Guernsey Press Co. Ltd.

Busby, D., & Rutland, C. (2019). *The Horse a Natural History*. London: Ivy Press.

Clutton-Brock, J. (1987). *A Natural History of Domesticated Mammals*. London: British Museum (Natural History).

Dawkins, R. (2016). *The Selfish Gene*. Oxford University Press.

Gaunitz, C. *et al.* (2018). Ancient genomes revisit the ancestry of domestic and Przewalski's horses. *Science*, *360*, 6384.

Goodwin, D. (2007). Horse behaviour: Evolution, domestication and feralisation. In N. Waran (Ed.), *The Welfare of Horses: Animal Welfare* (pp. 1–18). Dordrecht: Springer.

Librado, P., Fages, A., Gaunitz, C., Leonardi, M., Wagner, S., Khan, N., Hanghøj, K, Alquraishi, S. A., Alfarhan, A. H., Al-Rasheid, K. A., Der Sarkissian, C., Schubert, M., & Orlando, L. (2016). The evolutionary origin and genetic makeup of domestic horses. *Genetics*, *204*(2), 423–434.

Orlando, L. (2020). The evolutionary and historical foundation of the modern horse: Lessons from ancient genomics. *Annual Review of Genetics*, *54*(1), 563–581.

Puruggan, M. D. (2022). What is domestication? *Trends in Ecology & Evolution*, *37*, 8.

Chapter 2

Equine ethology

The study of animal behaviour

When studying animal behaviour there are a number of questions researchers will ask about the activities they observe, known as Tinbergen's four questions. Nikolaas Tinbergen was a Dutch biologist and ornithologist, renowned for his foundational work in the field of ethology. Tinbergen, Karl von Frisch and Konrad Lorenz were awarded the Nobel Prize in Physiology or Medicine in 1973 for their research into patterns of animal behaviour. Tinbergen's research greatly advanced our understanding of animal behaviour through his famous framework of "four questions", which look at behaviour through the lenses of function, evolution, mechanism and development (Bateson and Laland, 2013).

1. *Function*: What is the purpose of the behaviour? How does engaging in this behaviour enhance the animal's fitness, specifically its survival and ability to reproduce?
2. *Evolution*: How did the behaviour evolve over time? What role has natural selection played in shaping this behaviour throughout the history of the species?
3. *Mechanism*: How does the behaviour operate? What specific stimuli trigger the behaviour, and what physiological processes underlie its execution?
4. *Development*: How does the behaviour develop during the individual's lifetime? How are the expression and modification of this behaviour influenced by experience and learning?

In this chapter we explore equine ethology predominantly by focusing on the first of Tinbergen's questions – what is the behaviour for, why is the horse performing the behaviour and how does it increase the animal's fitness? But first let's define what we mean by ethology.

Ethology – what is it and why is it important?

Ethology is the study of the function and evolution of an animal's behaviour in its natural environment. It helps to explain how that behaviour contributes to the animal's success in terms of survival and reproduction. As discussed in chapter 1, there are no truly wild horses

DOI: 10.1201/9781003581390-3

alive today and it is not possible to fully understand the ethology, or natural behaviour, of any wild animal in a captive environment, be that a zoo or a livery yard. As a result, the study of the horse's ethology relies on observations made using free-roaming populations to investigate how those herds function in their natural environment. It's worth reflecting on how easily previously domesticated horses can thrive in conditions that are minimally managed, or not at all. In fact, studies suggest that although the physical characteristics of domestic horses may have altered from those of their wild ancestors, little of their species-specific behaviour has changed. The horse as a species remains physically and mentally adapted to a life on open plains, on mountains or in woodlands, with a behavioural repertoire that is relatively unaltered despite its evolutionary journey.

The principal role of that behavioural repertoire is the integration of the horse with their physical and social environment in order to maintain homeostasis. Homeostasis is the balance or equilibrium needed to maintain a stable internal environment, allowing the body to survive and function effectively while adjusting to changing external conditions. In essence, homeostasis represents the optimum physiological basis of wellbeing. If we go one step further, we can say that ethological homeostasis is the recognition that the body uses behavioural processes to attain that physiological stability. In other words, the horse is innately motivated to behave in certain ways in order to maintain their physiological wellbeing.

A detailed understanding of the behavioural processes of these free-roaming horses and therefore what is considered normal behaviour is essential for the wellbeing and welfare of horses managed in a domestic setting (Waring, 2003; Mills and McDonnell, 2005). The ability to appreciate the horse's natural ethology and be able to distinguish between normal (albeit unwanted by human caregivers) behaviours and the abnormal behaviour developed to cope with a human-engineered environment is crucial for anybody involved in modern horse management. We'll delve deeper into equine welfare and its assessment in Part 3. For now, let's take a look at the equid ethogram.

The equid ethogram

An ethogram is used by behavioural scientists to describe the range of behaviours displayed by a species in a specific environment. The behaviours observed can vary depending on the circumstances in which the animal is being studied. Free-roaming animals may demonstrate different behaviours to those living in captivity due to the availability, or lack, of resources within that environment and the opportunities to perform certain behaviours. However, before we can distinguish between natural behaviours (i.e. those observed in free-roaming populations) and those developed to cope with domestication, we need to identify what "normal" horse behaviour looks like. To do this we use the equid ethogram.

An ethogram is a comprehensive catalogue of the behaviour of a species that provides insight into their behavioural needs that culminate in homeostasis. We will explore the equid ethogram under five general categories:

- Maintenance behaviours – these are the basic, fundamental behaviours that all horses need to perform on a daily basis in order to stay alive and well.
- Socialisation and affiliative behaviours – horses are highly social animals and will generally choose affiliative and avoidance behaviour over conflict.
- The role of play – horses play throughout their lives providing both physical development and learning.
- Sexual behaviour in horses – natural courtship and mating behaviour.
- Developmental behaviour in foals – the crucial time period after birth up to weaning.

Maintenance behaviours

Maintenance is a term used to describe a major category of behaviours directed towards ensuring a homeostatic state as well as establishing the psychological wellbeing of the horse. These are the behaviours that are critical to life and you'd be right in thinking that they include eating and drinking. However, ethological studies show that it is a lot more complicated than that, and as horse owners, trainers, breeders, vets and therapists we need to understand that maintenance behaviours go way beyond the basic needs for life. The following is a prioritised list of maintenance behaviours identified through ethological studies.

- Safety – as a social species horses rely on others for safety
- Ingestion – grazing and drinking
- Body care – evacuation, grooming, stretching and temperature regulation
- Rest – drowsing, resting and sleeping
- Motion – movement, ranging and playing
- Exploration – investigative activities

Safety

Safety and ingestion are probably the most fundamental and therefore prioritised maintenance behaviours in terms of survival. Safety means the ability to respond to actual or perceived environmental threats. As a prey species, group living is an important survival strategy as it increases the likelihood of detecting approaching predators and also reduces the probability of any particular individual being caught and eaten. Horses are social animals and whilst preferring to associate with others of their own kind, they will accept other species as companions and respond to the warning signals of those other species – including humans! This is worth remembering when we're feeling nervous, anxious or worried prior to interacting with our horses. We'll explore the horse-human relationship further in chapter 3.

Like many social prey species, the horse's sensory system and social behaviour have adapted to facilitate early detection of approaching predators. The primary alert response in horses is an elevated head and neck with an intense orientation of the eyes and body. This posture in one horse sets off a ripple effect in others. If a perceived threat is recognised as being dangerous then the horses will take flight to create distance between themselves and the threat. If it is a false alarm, older, more experienced horses, having taken in information from a safe distance, will return to grazing within minutes and the younger horses will eventually follow their lead. Interestingly research has shown that horses, when confronted with a predator, demonstrate two main defence strategies depending on the environmental conditions and the nature of the predator. Environments with big cats as the most likely predator appear to be associated with the evolution of the rapid flight response which keeps the vulnerable head area as far away from the predator as possible, thus avoiding the kill bite to the throat. Whereas when canids are the main predators, often where the ground conditions are boggy and rapid flight has a high risk of injury, horses may be more likely to stand their ground and defend themselves using foreleg strikes. These strikes are more than capable of smashing the skull of a dog or wolf. There is also some evidence of breed differences in the type of strategy employed. Adoption of the flight strategy is characteristic of highly reactive breeds, such as the Arab who originates from hot arid areas with ground conditions conducive to galloping and where predation by big cats is probable. In contrast, Iberian horses, the Moroccan Barb and many British native ponies have been shown to preferentially employ a stand and fight strategy (Goodwin, 2007).

Ingestion

Horses in natural environments are predominantly grazers, and they will also browse on a wide range of shrubs, trees, leaves, stems, fruits, roots and seeds. Feral and free-roaming horses will spend up to 16 hours per day foraging in grasslands and other habitats containing a range of vegetation. Most grazing takes place during daylight hours, although the ratio of day to night grazing is affected by very hot weather and the presence of flies. Cold and wet spells in the winter also reduce grazing time. Seasonal and diurnal patterns of eating are characteristic of grazing behaviour in horses, with the total time spent grazing also influenced by grass quality and availability. The diet of New Forest ponies has been shown to be seasonal with grass constituting 60% of their diet in the summer, dropping to 30% in the winter when the ponies are more likely to be found foraging in the sheltered habitats amongst the gorse and deciduous woodland. Researchers investigating the seasonality of free-roaming horses found that both Przewalski's horses and Shetland ponies were able to adjust their energy budget to accommodate predictable changes in forage quality between the seasons. What this means is that despite their evolutionary journey, horses have continued to adapt seasonally to environmental conditions via fluctuations in their metabolic rate. It also highlights why there is a need to adjust our domesticated feeding practices at different times of the year in line with these metabolic fluctuations. Another potential remnant from an ancient past is the propensity of some horses to debark the trunks of trees and chew wood. This is generally considered by ethologists as vestigial browsing behaviour and is a good example of "normal" albeit sometimes unwanted behaviour, especially if it's your trees or fence posts that are bearing the brunt! While chewing, horses will lift their heads to check for threats, and naturally living horses generally eat slowly and carefully using their sense of taste and smell to select specific herbage and reject unfamiliar or unpleasant items. The structure and function of the horse's stomach requires the animal to keep moving during an almost continuous intake of small amounts of food, and this explains why grazing horses will only crop a couple of mouthfuls before taking a step forward and repeating the process.

In terms of drinking, horses ingest water in the form of moisture both in and on forage as well as from streams, pools and puddles. During the winter, they can also ingest water in the form of snow, frost and ice. In the natural environment, horses generally drink infrequently in a 24-hour period and many may only drink once or twice a day, but when they do drink will typically consume a large quantity. The frequency of drinking and the volume of water required varies with age, sex, weight, activity, reproductive status, health status, climate and diet. Lactation is an important factor in drinking requirements, with water intake increasing by 30%–60% above maintenance levels. Feral lactating mares have been observed to drink more frequently and therefore tend to remain closer to water sources. Unsurprisingly, water requirements for free-roaming horses have been shown to increase during dry conditions, with one study finding water intake up from approximately 23 litres per day to about 40 litres per day after short-term exposure to a hot humid environment. In Australia, where there can be large distances between good forage and water sources, drinking as little as once every 48 hours has been observed, with some horses recorded to have walked up to 55 km to watering holes. Given their dependence on water, especially in warm seasons, water distribution strongly influences the dispersal, space use and movement patterns of free-roaming horses. When water sources are more abundant, horses spend less time moving. In regions where it snows during winter, horses can eat snow for hydration, which enables foraging further away from normal water sources compared to summer months (Harvey *et al.*, 2022). Table 2.1 lists characteristics of types of ingestion behaviour.

There is a widely held belief that horses have an inbuilt mechanism that triggers the consumption of specific nutrients when a lack or deficiency is detected in the body. Whilst there are regulatory systems for water and sodium creating thirst and salt appetite, many studies show that horses given free access to salt licks are not able to self-regulate their intake to match their need and this is likely to be the same for any other mineral or nutrient.

Table 2.1 Ingestion Behaviour in Horses, Observed by Ethologists

Type of behaviour	Characteristics/purpose
Grazing	Ingesting grassy vegetation
Grazing while lying down	More commonly seen in foals, sometimes in uncomfortable adults that are reluctant to stand and sometimes seen in healthy adults
Pawing or digging while foraging/drinking	Removing snow or matted overgrowth that is covering the vegetation beneath. Uncovering roots or breaking ice.
Browsing	Ingesting woody plants
Coprophagy	Ingesting faeces – normal development behaviour of foals
Pica	Ingesting soil
Drinking	Horse may dig holes that fill with water or drink from hoof impressions that have filled with water. They can obtain considerable hydration by licking ice and ingesting snow.

Body care

The first of the maintenance behaviours that make up body care (Table 2.2) is eliminative or evacuation behaviour. Adult horses typically defecate six to 12 times per day depending on the availability of food. Normal urination occurs less often, around three to six times per day. Observations of feral populations suggest mares urinate and defecate *ad lib* with fairly random placement. In stallions, defecation is more deliberate with faecal accumulations positioned along commonly travelled routes or at important resources such as watering holes as a form of scent marking (McDonnell, 2003). Eliminative behaviour differs between domestic and free-roaming horses. Where domestic horses are confined to a fenced pasture they show an aversion to grazing near faeces which results in the development of separate grazing and toilet areas. In contrast, research on free-roaming horses demonstrated that, aside from marking behaviour, they eliminated indiscriminately throughout their home range. Interestingly, New Forest ponies do appear to have specific defecation areas within grasslands on the forest although their use is less obvious during the winter when grazing is limited. This suggests that whilst the use of a separate toilet area may be an adaptation to domestication it may also reflect a dynamic relationship between animal density and the availability of clean grazing (Goodwin, 2007).

The second maintenance behaviour under body care is stretching and grooming, both of which are exhibited by horses of all ages and play an important role in contributing to the wellbeing of the horse. The significance of these activities may easily be overlooked as they tend to be short and varied, but since they are repeated frequently throughout the day they are clearly a key part of self-care. Horses spend time grooming their own bodies, especially around the hip and flank and occasionally their limbs, both fore and hind. This self-grooming includes rolling, rubbing and scratching, as well as licking and nibbling the coat. The eye, face, nose and nostrils receive hygiene attention by rubbing the face against objects or up and down the side of a foreleg. Horses groom their crest by rubbing it back and forth beneath a branch or fence rail or against a tree. The same object can be used to scrub the hind end. Often considered an unwanted behaviour, rubbing or swaying the hind end against a tree or fence is a normal grooming behaviour designed to remove dead skin and scurf from the buttocks and tail

Table 2.2 Grooming/Coat Care Behaviours in Horses, Observed by Ethologists

Type of behaviour	Characteristics/purpose
Roll	Removal of loose hair, scratching and waterproofing. Rolling sites are often used repeatedly until they become barren and dusty. Most horses will roll from side to side.
Shake	Usually follows rolling and is used to dislodge debris from the coat.
Auto grooming (nibbling, biting, licking, scratching or rubbing)	May involve using parts of the body such as rubbing the head against a foreleg or using a hind leg to scratch an ear. Often requires the use of objects such as trees, fences and posts.
Mutual groom	Appears to serve as an aid for shedding, parasite control, stress alleviation and social bonding. Some research suggests the stimulation of natural grooming sites is especially calming.
Swash/swat/stamp	Swish tail to remove insects. Swing head against shoulder/ abdomen or flex chin to chest to dislodge flies. Flex/stamp limb to remove insects.

head. The motion causes the tail to be pressed into and across the skin of the dock, serving to wipe the hairless region which can accumulate skin scales, salt from sweat, grease and small faecal accumulations. Rolling and shaking are also forms of grooming in the horse. After one or more rolls from side to side the horse rises and shakes its whole body vigorously from head to tail. This action causes the skin to ripple over the entire body and dislodges any debris from the coat. To perform this grooming behaviour effectively, horses need the opportunity to roll on various substrates to allow for scratching, removing loose hair or creating a mud layer for waterproofing. Mutual grooming, although related to skin hygiene, is practised in a precise manner over certain limited sites indicating that physical comfort is derived from this activity and that social bonding is likely a primary motive. We'll discuss social and affiliative behaviour later in the chapter.

Pandiculation (Figure 2.1) is the term used to describe the systematic stretching the horse performs in a stationary position and is considered a somatic (mind-body connection) form of body care. Stretching is usually undertaken after rising and is a series of actions which may include flexion at the throat, arching of the neck, straightening of the back and extension of the hindlimbs or forelimbs. Such stretching appears to function as an expression of wellbeing, especially in foals, with healthy foals averaging around 40–50 pandiculations per day in various forms, both recumbent and upright (Fraser, 1997).

In terms of temperature regulation, we first need to understand the horse's thermoneutral zone. The thermoneutral zone defines the range of temperature in which the horse can regulate its body temperature with little or no energy expenditure. The general consensus is that a mature horse's average thermoneutral zone ranges from 5°C to 25°C although horses can adapt and live happily in temperatures as low as −15°C and as high as +30°C–40°C (Kentucky Equine Research, 2011). This means that the horse needs to take very little action to maintain body temperature when living in the thermoneutral zone. The temperature regulation behaviour we do see in free-roaming horses (Table 2.3) is the desire to seek out shaded areas or water to keep cool and the use of natural shelters, body positioning or companions to stay warm/dry in wet and windy conditions.

Figure 2.1 Konik horse in pandiculation. Photograph by Smeerjewegproducties.

Table 2.3 Temperature Regulation Behaviours in Horses, Observed by Ethologists

Type of behaviour	Characteristics/purpose
Backs to natural wind break – may stand alone or in a group	Provides shelter from cold winds and allows horse to stay warm.
Backs to weather either standing or grazing during heavy rain or strong winds	Serves to reduce area of body exposure and minimise heat loss.
Huddle – may be two or more in the group	Reduces area of body exposure, minimises heat loss and provides shelter.
Sunning – standing or lying in sunny warm areas	Provides warmth during cold weather.
Stand/rest in water	Horses have been observed to stand or rest in water for long periods or even lie down in knee high or shallower water in order to keep cool.

Rest

Horses require far less sleep than humans do, but it still plays an important role in their health and wellbeing by providing for both biological and physiological restoration. Sleep disorders or simply a lack of sleep may seriously impair a horse's physical activity and quality of life (Greening and McBride, 2022). Horses are crepuscular, meaning they are most active at dawn and dusk, and they have polyphasic sleep patterns, meaning they have intermittent periods of rest and sleep throughout a 24-hour period. In general horses require a total sleep time of three to five hours per day, and these are split up into five to seven episodes of 30–40 minutes in duration.

Horses rest in various ways (Table 2.4), with the most common form being drowsing, which can be achieved while lying or standing. The stay apparatus, a series of muscles and ligaments in the hind and fore limbs, lock the joints in place allowing the horse to stand with minimum muscular effort and thus drowse standing up. Horses spend around 2.5 hours per day recumbent with twice as much time spent in sternal recumbency (lying down but with the head elevated) compared to lateral recumbency (lying flat out). Healthy adult horses are unable to lie flat out for long periods of time before their respiratory function becomes impaired and therefore rarely spend more than 30 minutes continuously in lateral recumbency. This is not the case for foals or young horses who may be seen to spend many hours of the day sleeping on their sides at full stretch (Fraser, 1997).

Horses show both forms of deep sleep –slow-wave sleep (SWS) and rapid eye movement (REM). These two forms of deep sleep have been likened to the sleep of the mind and the sleep of the body. The REM phase of sleep in particular is crucial for both the psychological and biological wellbeing of the horse and they must lie either flat out, or in sternal recumbency with their muzzle on the ground, in order to achieve REM sleep. When standing and drowsing the head is down, eyes partly closed and the lower lip relaxed. One hind leg is usually flexed. In groups of horses, individuals take it in turns to lie down and sleep so there is always one "sentinel" horse who remains alert for dangers (Figure 2.2). Although it is unusual to see all members of a group of horses lying down at any one time they do idle simultaneously.

Figure 2.2 Horses in lateral and sternal recumbency, with one horse standing sentinel. Photograph by Kerry Neill.

Table 2.4 Rest Behaviours in Horses, Observed by Ethologists

Type of behaviour	Characteristics/purpose
Idling	Stationary standing for an extended time period with some limited limb shifting and positional changing.
Drowsing	Usually an upright stance with one hindlimb relaxed. Eyes closed, head lowered below the back, stay apparatus engaged, only light sleep is possible.
Resting	Can be standing or recumbent with evidence of wakefulness. Standing – head slightly lowered, eyes partly closed, often bearing weight on three legs, lips relaxed and ears rotated laterally. Sternal recumbency – rests on one side of chest with one foreleg and one hindleg underneath the body.
Sleeping	Usual sleep position is flat on the side with legs extended. In REM sleep rapid eye movements may be seen below the closed lids and there may also be minor movements of the leg or feet.

Motion

Locomotion is usually defined, in behavioural terms, as voluntary movement of the whole body and begins within the first hour of life. All healthy foals, by the end of their first day, will have a full complement of locomotion abilities – walk, run, jump and swim. Kinesis, or movement, can therefore be seen as a basic need in the ethological homeostasis of the horse. Genetically they are prepared to travel considerable distances on a daily basis to forage and will continue to do so even when the need to travel to find food has been removed. This kinetic need ensures a minimum level of activity required to keep the animal physically prepared for any necessary movement such as escaping a predator and is an important consideration in terms of welfare in domestic conditions.

Feral and free-roaming horses generally occupy a large area termed a home range. A home range contains the resources important for survival such as watering holes, suitable grazing areas and protected areas for the avoidance of insects and the extremes of weather. Studies show that home ranges vary in size dependent on the availability of these resources. Seasonal movements within home ranges indicate a form of rotational grazing system with some areas only being utilised at certain times of the year. A group of New Forest mares were reported to have a home range of between 82 and 1,020 hectares in size with movement to the shady areas in the summer increasing their home range over ten times compared to the winter range (Goodwin, 2007). This ability to freely move around their home range, including walking while grazing, reacting to threat and playing between herd members is a fundamental maintenance need for all horses.

In addition to the familiar locomotion behaviours of walk, trot, canter, gallop and jump, ethologists have also observed the following normal movement behaviours in free-roaming herds:

- Trek – where two or more animals move a distance together typically one following the other.
- Stampede – a group of animals running as a unit at high speed.
- Swim – horses wading and swimming in deep watering sites.

Exploration

Horses show a strong motivation to explore and investigate their environments. This exploratory behaviour (Table 2.5) is most obvious in young animals but is seen in all groups of horses as they roam their surroundings. Once the environment is familiar the motivation to explore subsides but will reappear whenever there is a change such as a change of field, the addition of a new herd mate or a novel object. Useful information and experience is gained through adaptive exploration. Inquisitiveness and the resulting exploratory behaviour in horses is a healthy sign and plays a considerable role in learning and adaptability, something that behaviourists and trainers can tap into (more on that in Part 2). Trial-and-error experience is of particular value as the horses learn about their own abilities, limitations and resource holding potential. They do this by investigating them through social interaction and interactions with their environment. Horses will explore any new environment they find themselves in and pay most initial attention to the boundaries before exploring the interior. When encountering novel objects, odours or a newborn addition to the group horses will typically use all their various senses to investigate it. At first this generally involves tentatively looking and smelling and then gradually progresses to mouthing, pawing and tasting.

Time budgets

A time or activity budget describes how a horse's day is broken down into the different behaviours on its ethogram. The time budgets of free-roaming horses have been well researched and show that their daily rhythm of behaviours is linked to sunrise and sunset, when they are most active, and influenced by the ambient temperature, season, availability of forage and the presence of insects. Typically, grazing occurs predominantly in two phases, the first in the early morning and the second in late afternoon. Longer periods of time are spent feeding during winter than summer and lactating mares spend more time feeding than non-lactating mares. Resting occurs generally during the middle of the day with recumbent rest most common between 2 am and dawn. Movement unrelated to feeding makes up a relatively small amount of the time budget, with movement time related to the distance between prime feeding areas and water sources. However, it is worth remembering that horses are constantly moving while they're eating, taking one or two steps for every mouthful of grass. Stallions have been found to spend more time moving and less time feeding than mares (Harvey *et al.*, 2022).

Table 2.5 Exploratory Behaviour in Horses, Observed by Ethologists

Type of behaviour	Characteristics/purpose
Sniff/lick	Draw in air through the nostrils or touch the object with tongue in order to investigate the odour, texture, shape, taste and size of an object or feature. Often seen when meeting a newborn addition to the herd.
Nibble/mouthing	Appears to be a means of moving an object on the ground. Object may be held between the lips or front teeth and then dragged forward and back and side to side.
Paw	Toe dragging movement as if digging. This movement is repeated several times and may have direct contact with the object or feature being investigated, e.g. water.

Social structure

Horses are highly social animals (van Dierendonck, 2006) and when free roaming will live in relatively stable units called bands, of which there are two main types – a harem or natal band and a bachelor band. Harems usually consist of one or more stallions, several often unrelated mares and their foals as well as yearlings and two-year-old horses of both sexes that are yet to disperse. The average size of a harem tends to range between four and 13 horses, although some have been reported as large as 35 (Harvey *et al.*, 2022). Fillies remain with their mothers for two or more years before changing groups, often joining older mares with foals. Juvenile colts and stallions without their own harem gather in bachelor bands. The majority of bachelor band members are yearling stallions who have left their natal groups and young adult stallions yet to secure a harem of their own. Sometimes a displaced older stallion will join a bachelor group, or they may choose to live alone. Bachelor bands can range from two to 17 horses and they are generally less stable than the family groups. The size of bands is influenced by a range of factors including food and water availability and the risk of predation. They tend to be larger in more open environments, especially when resources are plentiful. Severe weather conditions in winter and food limitation during dry seasons cause bands to split into smaller groups where they roam in larger and overlapping areas. Interestingly, domestic horses running together in large areas tend to split themselves into bands even when no stallion is present; these bands often aggregate into larger herds around water sources or preferred grazing sites. As discussed in chapter 1, horses are not normally territorial and stallions will generally only defend their mares rather than the space they inhabit.

Researchers have found that it is often circumstances that influence the way horses live socially, and this flexibility means that although their affinity for social life is strong, horses are not fixed in the way it should be organised, although there are a few common factors, one of which appears to be family. In the New Forest, even after weaning, youngsters tend to stay with or around their mothers. They do become more independent and exploratory but will still tend to follow the dam if she makes off in a purposeful manner. In the wider band, any member may initiate movement of the group, usually in the context of resource-related activity including moving to and from locations of eating, drinking, resting and sheltering.

Part of the role taken by the stallion in a band is to herd the members, working from the back to keep the mares and juveniles in a defendable group. Stallions play a central role in the structuring of bands by controlling the movements of harem members and driving maturing colts out of the group. In this way they minimise the risk of extra-group fertilisations and the straying of harem mares; however, researchers have observed determined mares finding ways and means of mating with an alternative choice, and as in other societies, not all offspring are related to the breeding male in the family. Stallions have been observed to spend more time loafing and grooming with favourite mares, even when not in season, and will readily play with any foals, fillies or colts (Ransom and Kaczensky, 2016). The presence of a stallion has a significant influence on the group, and interactions among individuals and the social structure in groups without a stallion may be significantly altered from the natural state (Sigurjónsdóttir *et al.*, 2003). The significant differences between the structure of a free-roaming band and a captive, managed group typically without a stallion but including one or more castrated males are factors to be aware of in understanding the behaviour of domestic groups and individuals.

Another common factor in the social structure of horses is friendship. We often see examples of these friendships in a managed setting with preferred or long-term companions swishing flies off each other, hanging out and mutually grooming. Studies in feral populations have also shown that in groups grazing freely, hanging out or dozing the same horses are often seen close together, and it is the ones they stay closest to that they groom most often. Clearly, horses are capable of forming close social bonds and a lack of social contact is likely to be one of the most serious stressors for the managed horse today. Besides grazing, resting and grooming together, the formation of mutually supportive pair bonds plays an important role in social support

Table 2.6 Features of Social Behaviour That Contribute to Group Stability

Social behaviour	Group benefit
Eating, resting, sheltering together	Social facilitation – bonding
Nose to nose greeting, mutual grooming	Social exchanges – bonding, reducing conflict
Bunching to shelter	Protection
Following the group in a particular direction	Draws all members into a common shift of place
Initiate movement, activity or direction	Facilitates use of resources

(Table 2.6), particularly in difficult situations. Pair bonds can also be established with other species such as donkeys and goats, and this tendency may also be the foundation for horse-human bonds which extend beyond a utilitarian relationship. In essence, friendship, the need for contact and a biological preparedness to follow an attachment figure shapes the behaviour of social animals, and it is no coincidence that this facilitates our attempts to manage and modify their behaviour by leveraging this natural understanding of friendship, contact and cooperation.

Affiliative behaviour

The social behaviour of the horse functions to minimise conflict: overt aggression in free-roaming horse bands is relatively rare, not surprising when we recognise that such behaviour expends valuable energy better used to enhance fitness to survive and reproduce. Horses' social behaviour also ensures stability amongst band members. Social affiliation is essential to cooperative and collaborative behaviour, and, unlike other mammals where the sexes only maintain bonds during the mating season, stallions stay with their harem throughout the year under natural conditions. The two main affiliative behaviours demonstrated by horses are mutual grooming (also known as allogrooming) and social play (van Dierendonck *et al.*, 2009). Allogrooming strengthens bonds, reduces social tension between group members and provides reassurance following conflict. It can have a calming and stress-reducing effect, reflected in lower heart rates and cortisol concentrations in the peripheral circulation particularly when performed at a site close to the withers. Play within peer groups has a vital role in the development of the horse with up to 75% of the kinetic activity of foals devoted to play, but it also continues to have an important role into maturity, particularly in males. Play is usually characterised as activities having no immediate purpose or function; however, social play is different. It involves learning sets of rules which inform the relationships within the band and build both social and communication skills. Association with humans is a special form of affiliation, and associative tendencies in horses are increasingly being recognised as a manifestation of their conscious choice of, and need for, company. Facilitating the development and maintenance of these relationships undoubtedly benefits the welfare of horses in the managed environment.

Avoidance and agonistic behaviour

The operation of an avoidance system in the social behaviour of horses is a vital response in agonistic situations. Agonistic behaviours are those which cause, threaten to cause or seek to reduce physical damage during direct encounters between members of the same species. The aim of an avoidance system is to increase the distance between individuals through aggressive or defensive interactions – biting, kicking, chasing, barging. In general, offensive behaviours are performed with the head and forelimb, while threats with the hindquarters can be used both offensively and defensively (Table 2.7). True aggressive behaviours are avoided in favour of cooperation among the social group because aggression uses up valuable energy, presents a risk to health due to injury and serves to separate individuals and weaken crucial group cohesion. Thus, avoidance is the chief social tactic involved in maintaining social stability in a band. However, for that to be effective it requires sufficient space to get out of range of an aggressor. If the available area is too small to allow an individual to get outside the perimeter of an aggressor's space when they are being targeted, then an avoidance tactic cannot operate. This explains why aggression is far more frequently observed in the managed environment which tends to

Table 2.7 Examples of Agonistic Behaviour Observed in Bachelor Bands

Type of behaviour	Characteristics/purpose
Approach	Forward movement towards another in a straight or curving line – any gait or speed. Head may be elevated with ears forward or lowered with ears pinned back. May be followed by retreat, olfactory investigation or fighting.
Arched neck threat	Neck tightly flexed with the muzzle drawn down toward the chest. Observed during close aggressive encounters and ritualised interactions. Arched neck threats are a component of most male-male interactions.
Retreat	Movement that maintains or increases an individual's distance from an approaching individual. Can be at any gait but typically at the trot. Head is usually low with ears turned back.
Bite	Opening and rapid closing of the jaws with teeth grasping the flesh of another individual. Ears pinned and lips retracted. Biting at each other's legs is common among bachelor stallions and is the primary fighting tactic of horses.
Bite threat	Similar to a bite except no contact is made. The neck is stretched and the ears pinned back as the head swings toward the target. The miss appears deliberate as opposed to accidental or successfully evaded and acts as a warning to maintain distance.
Chase	One individual pursues another usually at a gallop in an apparent attempt to overtake, direct the movement or catch up. Chaser typically pins the ears, exposes the teeth and bites at the pursued horse's rump and tail. Chasing is usually a part of fight sequences.
Circling	Two individuals closely beside one another, head to tail, pivot in circles usually biting at each other's flanks, scrotum, rump/hind legs. With prolonged circling may progress lower to the ground until they reach a kneeling position where they continue to bite or nip one another.
Defecate over/ elimination marking	Defecation on faecal piles – sniff, step forward, defecate, pivot or back up, sniff again. Stallions appear to compete to be last to defecate on a pile.
Ears laid back/pinned	Ears pressed flat back against the head and neck. Typically associated with intense aggressive interaction. Ears are the most compressed at the most intense point of an aggressive interaction.

(Continued)

Table 2.7 (Continued)

Type of behaviour	Characteristics/purpose
Head threat/head swing	Head lowered with the ears pinned back, neck stretched or extended toward the target. Often lips are pursed.
Kick	One or both hind legs lift off the ground and rapidly extend backwards toward another individual with apparent intent to make contact. Neck is often lowered. Common for two individuals to simultaneously kick at each other's hindquarters.
Kick threat	Similar to kick but without sufficient extension or force to make contact with the target. Hind leg(s) lift slightly off the ground and under the body in tense readiness usually with no subsequent backward extension of the hind leg(s). Individual may turn their rump and back up to target. Tail may lash and may vocalise a harsh squeal. Appears carefully aimed and seems to serve as a warning, helps to maintain distance between individuals.
Kneeling	Drop to one or both knees by one or both horses engaged in face to face combat or circling with mutual biting or nipping repeatedly at the knees, head and shoulder. May be a defensive action to protect the forelegs from bites of the opponent.
Lunge	Swift forward thrust or charge from close range toward another individual. Most often displayed with a bite threat with ears pinned.
Nip	Similar to bite but with the mouth less widely opened and teeth closing on only a small piece of flesh. Nipping is seen during play fighting, mutual grooming and moderate to serious aggressive interactions.
Olfactory/social investigation	Involves sniffing various parts of another's head/body. Typically begins after two horses approached one another nose to nose. After mutually sniffing the face, typically one individual works their way along the other's body length. During investigation it is common for one or both to squeal, snort, kick threat, strike threat or bite threat.
Push	Pressing of head, neck, shoulder, chest, body or rump against another in an apparent attempt to displace or pin the target against an object.
Rear	Two individuals will rear in close proximity and attempt to bite one another on the head and neck or strike with the forelegs. Rearing of one or both stallions during sparring often appears to be a means of demonstrating superior height.
Stomp	One foreleg is raised and lowered, sharply and firmly striking the ground, usually repeatedly. Most commonly seen during posturing and ritualised interactive sequences.
Strike	One or both forelegs are rapidly extended forward to contact another individual, while the hind legs remain in place. The strike is typically associated with arched neck threat and posturing. A stallion may also strike when rearing. The strike is often accompanied by a squeal or snort. The strike can also be observed during ritualised interactions, with the usual point of impact being the ground.
Strike threat	An abbreviated strike in which a foot is lifted off the ground in a gesture that mimics the preparation for an actual strike. However, no forward motion with the leg or contact with a target is actually made. The strike threat is often seen as part of ritualised interactions between stallions and is frequently accompanied by a loud squeal or snort.

(Adapted from McDonnell and Haviland, 1995).

be more contained and less spacious than a free-roaming population's range. Most of us will be familiar with a yard where there appears to be a horse who is the "boss" and is always first at the gate or water trough and drives off others with threats of bites or kicks. Equally there is often an apparently lower ranked horse that gets "bullied" by the others and tends to loiter at the back during feed times. This can lead to the mistaken belief that within a group of horses the biggest, strongest, more aggressive horse is the "natural" leader. However, studies show this is not the case; rather it is more likely the result of competition for limited resources. Free-roaming horses do not have small piles of food to argue over and do not have to jostle for position at the gate or for the attention of owners, who may or may not have treats in their pockets!

The other problem associated with the managed environment and particularly in livery yards is the regular and unnatural changing of the membership of the social group. This instability often results in high levels of aggression and resultant injury (Busby and Rutland, 2019).

In natural social groups agonistic behaviours are most likely to feature during interactions among stallions. These interactions can range from peaceful and cooperative (a readily negotiated deference) to ongoing sparring which, although robust, rarely appears to result in injury. Incursions into a band by other stallions are usually prevented by posturing which effectively signals the stallion's relative fighting ability. There is a large range of ritualised behaviours that stallions use to assess each other's strength and avoid unnecessary direct contact. An ethogram of agonistic behaviours has been developed and although these were identified in bachelor stallion bands you may well recognise many of them from encounters between horses in general daily life. It's also worth mentioning that some of the research was on pony stallions living in pasture bands of three to 20 mature stallions in fields (they used 0.3–0.5 acres per pony stallion) with multiple watering sites and with very few incidences of even minor injuries. This was despite some spectacular displays of agonistic interaction when they were first introduced and is also consistent with studies of free-roaming Przewalski's horse stallions. Perhaps it's time to rethink our approach of isolating stallions in the managed environment in order to avoid fighting and the potential for injury.

Many of the agonistic behaviours of bachelor stallions are the same behaviours observed among juveniles of both sexes during play. In this context, they also serve a range of adaptive functions including musculoskeletal development, cardiovascular fitness and honing skills, particularly in terms of reproduction and survival.

Resource holding potential

Resource holding potential (RHP) is a fundamental concept in equine ethology, crucial for understanding the dynamics of social behaviour among horses. RHP refers to an individual's ability to acquire and retain essential resources. It's influenced by various factors including physical strength, age, size and even mental or emotional resilience.

In a settled band in the wild, access to resources such as food, water, and mating opportunities are regulated through individual RHP. In terms of reproduction, a stallion with a high RHP is more likely to control a larger territory and attract a greater number of mares, ensuring his genes are well represented in future generations. RHP not only determines the outcome of actual resource holding contests, it also influences their frequency and intensity, or even whether to engage in a contest at all: RHP behaviour reflects an innate cost-benefit analysis, where the potential costs of injury or death are weighed against the benefits of gaining or maintaining access to a resource.

RHP applies broadly to all types of resources that are valuable to horses, including not just territorial claims or mating opportunities, but also subtler preferences such as access to the sunniest spots for warmth or the best grazing areas. For example, during colder weather, the sunniest patches in a pasture might become highly desirable locations. Horses with higher RHP may assert their access these spots, which not only provide warmth but may also offer protection against harsh weather conditions. Similarly, in a grazing scenario, the areas with the most nutritious grass will attract more competition, and those horses with higher RHP are more likely to have uninterrupted access to these areas. From a practical standpoint, once the need that increased one horse's RHP is satisfied (e.g. they have warmed up), they no longer have such a strong motivation to defend this resource. They might simply walk away, making the spot available. This shift allows another horse, perhaps one who has been waiting and whose RHP or motivation to access the sunny spot is now greater, to take over without confrontation.

Crucially, though, this "resource management system" requires a group of horses to remain settled and consistent in terms of its members, and we know that typical domestic horse management involves a lot of group disturbance, whether that's a horse being sold or moving yard, a yard owner deciding to rearrange who is turned out in which field or even a horse going off to a show for the day, or horses being brought in at night and turned out again in the morning; even this can have an unsettling effect on such an affiliative species where togetherness is key. Safe and conflict-free management of groups of domestic horses requires an understanding of these dynamics, helping to create environments where resource competition is minimised by ensuring multiple sources of valuable resources and thus reducing conflict and enhancing welfare for all individuals in the group.

The role of play

Horses play throughout their lives. While play is most conspicuous in foals it also commonly occurs in young horses and bachelor stallions and least often in adult mares. Play is energetically costly, but research has shown that individuals that played more had better survival and body condition as yearlings even when they were weaned earlier (Cameron *et al.*, 2008). We already know from scientific study that early exercise influences physical development in horses, including tendon strength and flexibility as well as skeletal and muscular growth. But it goes beyond just the physical. Although play movements are innate they are not perfectly performed on the first occasion and some learning and experimentation is required. Play behaviour can be a solitary activity as well as a socially interactive behaviour (Figure 2.3), and the capability of each animal to thrive as an individual and within the group depends on the developments that play has encouraged. There may be some differences in the quantity and quality of play dependent on the social environment in which a foal develops. In free-roaming harem and bachelor groups there seems to be considerably more play in foals, yearlings, young adults and some mature males than found in mare and foal pairs or herds. Young foals and yearlings appear to mutually stimulate play, increasing the amount of locomotor social play in the younger foals and likely increasing the distance young foals play away from the dam. Harem stallions also engage in play with the foals and yearlings and contribute to an increase in the amount of play compared to mare and foal only groups. Similarly, bachelor stallions will engage in play fights and chases with yearlings and two-year-olds, further increasing the amount of play activity. It begs the question, what is the cost to the modern horse kept in managed environments that restrict this access to sources of play? In addition to playmates, play frequently appears to be stimulated by novel conditions or stimuli and is often interspersed with investigation and exploration. New experiences such as changes in weather, encountering unfamiliar objects and the movement of the group to areas not yet explored are associated with an increased frequency and duration of play behaviour. Play involving two or more individuals often involves agonistic behaviours; however, an important observable difference between play and more serious confrontations is that in true aggressive encounters the ears are generally

Table 2.8 Play Behaviour in Horses, Observed by Ethologists

Type of behaviour	Characteristics
Object play Involves contact and manipulation of an object. It may be an environmental object or the mane, tail or other body part of a herd mate or even an animal of another species.	Nibbling is typically one of the first play responses associated with investigation of an object. Sniff/lick – the tongue makes contact with the object. This may be to investigate the odour, texture, shape, taste and size. Chew – the object is picked up, may be shaken or pulled from side to side or up and down, carried about and then dropped or tossed. Paw – the object is touched by a foreleg in a digging motion in order to move it around. Water play typically includes pawing. Circle – horse moves in a path around an object at any gait. Mostly seen in foals and adolescents, less so in mature animals.
Sexual play behaviour Frequent in foals and young adolescents of both sexes as well as bachelor stallions.	Tease – one animal sniffs/nuzzles the body of another in the manner of a stallion investigating a mare. Mounts – one individual raises their chest and forelegs onto the back of a herd mate (same or opposite sex).
Locomotor play Any sequence that is performed while in motion.	Frolic – fore and hind legs simultaneously propel off the ground. May take off from a stationary position to an instant gallop. Run – spontaneous burst of motion with no apparent destination or threat to escape. Chase – an animal is pursued at trot, canter or gallop with the apparent intention of catching up and overtaking. The chaser often attempts to nip at or bump into the pursued partner. Also includes bucking, jumping, leaping, prancing.
Play fighting Involves sequences similar to serious adult fighting but with partners alternating offensive and defensive roles.	Nipping/biting to head, neck or chest. Neck grasp – mane and neck are held at the crest and sometimes moved back and forth. Foreleg nip/bite causing partner to buckle at the knees and partially drop to the ground in an attempt to evade contact. Mutual hindlimb nipping/biting. Bite or grasp of rump – this is usually when rugs get ripped! Push – head, neck, shoulders, chest body or rump is pressed against a partner in an attempt to displace them.

pinned back against the head. Ethologists have identified four distinct categories of play behaviour – object, sexual, locomotor and play fighting (Table 2.8) (McDonnell and Poulin, 2002).

Hopefully by now it is clear that play behaviour has a major role in the behavioural, social and physical development of foals and mature horses alike. I argue that play is in fact a basic requisite in the conditions of welfare, including as it relates to the development of today's sport and riding horses. The question for those of us involved in riding and training is how best to incorporate this motivation to play into the athletic and cognitive stimulation of our horses.

Figure 2.3 Horses play biting each other's faces. Photograph by SannePhoto.

Sexual behaviour in horses

Under natural conditions, reproduction in horses is seasonal. The hormones that control ovulation are light-controlled, explaining why mares are generally only fertile from March to September when daytime duration is longer than, or equal to, the night. Equally testosterone, the male sex hormone, is also influenced by daylight which is why during winter months stallions are fertile but don't normally demonstrate an urgent desire to mate. During the breeding season mares will typically complete 12 oestrus cycles lasting 21 days with sexual receptiveness (oestrus) lasting four to five days. Courtship under natural conditions starts several days before oestrus and begins with the stallion actively discouraging contact between the mare and other males. Interestingly, studies have shown that harem stallions appear to be more attracted to older, fully grown mares than the younger females. In fact, there was a significant correlation between foaling rate and age, meaning that the older, more experienced mares had higher foaling rates. Several other studies found that these older females would also restrict access to the harem stallion by chasing away younger mares or placing themselves between the youngster and the stallion. This suggests that the higher foaling rates among such mares may be the result of both suitability-related mate choice by stallions and direct female competition for mating opportunities (Powell, 2008).

As the mare enters full oestrus the stallion remains close and begins to test her readiness to mate by licking and nuzzling. This stimulation causes the mare to expel urine and adopt a standing position with back legs straddled and tail raised and arched to one side. At this the stallion's advances may be rejected with much squealing, tail lashing and stamping. The stallion will continue to assess the mare's willingness to mate by nudging and nibbling at her flanks, hindquarters and hind legs. Only sexually receptive mares will stand for mounting, whilst unreceptive mares will kick out if the stallion attempts to mount before they are ready. When full oestrus is reached and successful mating has taken place the pair will remain together for up to several days and mate repeatedly. This natural form of sexual bonding ensures the optimum conditions for fertilisation to occur. Following a gestation period of 327–367 days, mares normally give birth over spring and summer with a single foal. Mature females are capable of foaling every year, but most commonly they raise one foal every two years.

In the human-engineered breeding environment, stallion management and the restraint of mares during in-hand serving are areas of welfare concern for many ethologists. Through studying feral populations, we know that a harem stallion and his mares are together and interacting continually throughout the year and not just at breeding time. When the mare is in oestrus the frequency of interactions increases, with some researchers suggesting as many as hundreds of communications and precopulatory "teasing" sequences as often as every few minutes. By contrast, in managed breeding programmes mostly carried out "in hand", stallions and mares are typically allowed little to no contact other than brief and directed precopulatory interaction immediately prior to mating. So, it's no surprise that one of the most common problems with stallions in domestic breeding programmes is failure to show normal sexual interest in mares. In addition, studies have revealed that harem stallions show different behaviour towards mares of different ages. Mature mares are preferentially guarded, actively courted and mated by the harem stallion. This indicates that stallions recognise and alter their sexual behaviour depending on the age and experience of the mares in their group. Perhaps this helps explain why, in domestic hand-bred situations, there are times when a stallion may not accept or perform as well with a particular mare or mares.

Observations of free-roaming herds have also demonstrated that mares play a far more important role in terms of stimulation of the male and actual time of breeding than is generally assumed in modern stud farms. In harems, the mare in oestrus is very obvious by her increased movement and approaches to the stallion. Although in the early stages these approaches are solicitous they often result in rejection just before the stallion mounts. During the ongoing interactions over the days of oestrus there is usually a pattern of apparently increasing attraction and receptivity of the mare. Almost 88% of stallion and mare precopulatory interactions that lead to successful mating are initiated by the mare approaching the stallion rather than the other way around (McDonnell, 2000). It is the continued and persistent teasing by the stallion that eventually leads to the mare tolerating being mounted. During in-hand breeding there are limited opportunities for precopulatory contact which can prevent the mare from making the natural transition from resistant to receptive behaviour. This is likely to contribute to the high frequency of resistance to mounting and can lead to the use of restraints which further interfere with normal interactive behaviour and receptive posturing. Again, once you are familiar with natural reproductive behaviour it becomes clearer that when handlers direct the excited breeding stallion to the mare from the rear for immediate mounting, they are not allowing the natural nose to nose approach and systematic precopulatory investigation which precedes full oestrus and the mare's acceptance of the stallion.

Another consideration is the common practice in large breeding facilities of stabling stallions together in areas separate from mares. This effectively mimics the bachelor groups we see in feral populations. Research has shown that there are important behavioural and physiological differences resulting from harem or bachelor status. Harem status increases libido, testosterone levels and sperm production efficiency and quality, whilst bachelor status imparts changes in the opposite direction. What this means is that (intended) breeding stallions kept together, away from mares, show a suppressed reproductive function and lower sexual interest and arousal. By understanding the sexual behaviour of free-roaming horses you can appreciate how breeding outside the natural season, lack of or restricted opportunities to interact over time and separate housing of mares and stallions can all contribute to problems in the managed breeding setup and to compromised welfare of the horses therein.

Developmental behaviour in foals

Normal gestation in the horse is 11 months, meaning that the majority of foals are born in late spring or early summer when the grass is at its most nutritious. Foaling generally takes place late at night or early in the morning and many mares leave their social group to give birth. Almost immediately after birth, the foal begins to make struggling movements with the legs and tilt the head upwards. The period immediately after birth is a critical time for bonding between the mare and foal with the smell of the afterbirth and fluids seeming to assist in stimulating the mare's maternal reaction. Rising attempts begin soon after birth and involve coordination, balance and effort. Several failures to stand up usually occur with the foal falling back down each time. With successive trials improvements occur and soon the foal is able to rise to a fixed, partly upright posture. Once the foal is standing they begin to make random teat-seeking movements by exploring the underside of the dam. They instinctively seek dark areas and suck at the mare's legs and belly as they make their way eventually to the udder. This trial-and-error process in the foal's teat-seeking behaviour often occurs over an extended period of time – highly successful foals may complete it in 30 minutes while others may take several hours. To suckle the foal is required to adopt a parallel and opposite position with the mare and to extend and rotate the head and neck. To facilitate nursing, the mare flexes the opposite hind leg tilting the pelvis, and therefore the udder, towards the foal. Once a foal has sucked for the first time they will seek the mare's udder at regular intervals and with increasing confidence. Within two to three hours foals are able to stand, suckle and flee a predator with their mother if the need arose. Specific bonding between newborn and dam is established within a few hours and the mare thereafter rejects any foal other than her own. The foal's attitude is less specific and more opportunistic. A hungry foal may nurse from any mare, which might explain why some mares become aggressive towards other horses during the first few days after birth. This reaction is thought to be due to the foal's inability to immediately and completely imprint their own dam on their memory. There is a chance, at this early stage, that the foal may become confused and bond with another horse who was just being curious. The mare's aggressive behaviour avoids this risk, and she does not normally relax her protection until a few days after the foal is born when imprinting has had a chance to occur fully. In the managed breeding environment the presence or intervention of people at this crucial time can disturb the mare and the bonding process, including causing her to walk away from the foal thus delaying the time when the first suck is taken and interfering with imprinting.

Most maintenance behaviours are present in the foal within hours of birth including elimination, self and mutual grooming, investigation, mouthing of vegetation, social communication and locomotor sequences. Explosive locomotor and play behaviour is usually seen on the second day. During the first week foals spend most of the day resting and sleeping. Stretching becomes more frequent over days 2, 3 and 4 as the legs straighten and may occur standing or lying. Stretching assists the young foal in the athletic development of tendons, bones, joints and muscles. The most healthy foals have been observed to stretch as often as 50 times per day (Fraser, 1997). By the second week they are now spending only 50% of their time resting and are gradually beginning to eat grass. The young foal at first must spread and flex their forelegs to reach the grass and for a week or two until the neck has grown will be unable to walk and graze simultaneously. In these first few weeks of life the foal stays very close to the mare's side. Between one and two months of age they begin to make exploratory trips away from their mother, practising locomotory skills and beginning to play with other foals. From two to three months foals begin to spend much of their time associating with other foals and begin to form peer groups. Weaning takes place naturally at around nine to ten months, though some mares will continue to suckle their foal until shortly before the arrival of their next foal.

Foal handling

Mare-foal bonding occurs during the first hours after foaling, and maternal recognition and acceptance of the foal is particularly important in the first 12 hours (Rossdale, 2002). Researchers have found that early forced handling (stroking or bringing to the teat) did not appear to benefit the human-foal relationship (and it may compromise the relationship between mare and foal). In fact, bringing foals to the teat seemed to induce a later reluctance for human contact (Søndergaard and Jago, 2010). Even at two weeks old these foals kept their distance and continued to avoid the approach of people, unlike those where there had been no human intervention (Henry *et al.*, 2006). Other studies showed that foals that had been brought to the teat by humans for their first suckling were less playful and more reticent to leave their dams to explore (Hausberger *et al.*, 2007). In contrast, those foals who had received no human intervention were more independent, spent more time away from their mothers exploring the environment and more time interacting with social companions. Distance from mother and tendency to play are considered to be important characteristics of the mother-young bond and essential criteria for a secure base of attachment which is a necessary requirement for resilience later in life, including habituating to the human-managed environment and responding effectively to training interventions (Henry *et al.*, 2009).

There may be a number of different reasons why foals react negatively in response to early human contact. Firstly, the mares often demonstrate some nervousness to the proximity of people at this vulnerable time and this may affect the foal's perception of humans. Secondly, there may be an innate negative reaction to forced contact, with any interference during the first suckling impacting the establishment of the mare-foal bond. Finally, the reaction of foals to handling suggests that stroking as a pleasant experience is probably acquired later and is not present at birth. In reality, it is likely that a combination of all of these factors leads to the negative association between early forced handling and people. Therefore, wherever possible it is better to allow the mare-foal bond to become well established before initiating human intervention, especially during the crucial period of first suckling.

Weaning

In free-roaming herds, most foals are weaned between nine and ten months of age and certainly before the birth of the dam's next foal. Studies reveal that natural weaning induces no stress response in either animal and there are no overt signs of rejection from the dam either prior to or just after weaning (Henry *et al.*, 2020). The natural weaning process is a gradual one taking several months, after which the young horses remain with the natal group until they reach sexual maturity. At this point juvenile colts leave or are expelled by the stallion and many fillies join other existing harems or form new ones with males from bachelor groups. In the period between weaning and exit from the harem, the mare and foal retain a social bond and youngsters often develop pair bonds within their peer group. This provides continuity in the social life of the juveniles for some considerable time after weaning, which is a very different experience from that of many domestic foals. The early and abrupt separation of mare and foal, with the foal often being moved to a new environment, is a practice that presents a welfare concern. Artificial weaning has been associated with psychological, physical and nutritional stressors that are of welfare concern. It enforces the cessation of the mare-foal relationship earlier (usually around four to seven months of age) and more rapidly than would occur naturally, with the potential for long-term consequences in the development of the foal's social behaviour and their future trainability.

The factors that appear to influence the stress related to artificial weaning fall into three main categories. The first is the physical effects caused by the premature stopping of milk as the main nutritional source. This can lead to weight loss, growth suppression, possible gut sensitivity and problems with bone growth. The second is linked to changes in the social environment due to maternal deprivation, the use of social isolation as part of the weaning process and the placement of foals in groups with unfamiliar individuals. The final category is related to the environment to which the foal is moved and the effect this has on their ability to perform certain behaviours, for example, indoor versus outdoor, the amount of space available and the resources within that space. There appear to be fewer stress responses when foals are weaned gradually and allowed to have social contact either with other foals or with older horses. Ensuring the foal is still able to smell, see, touch and hear the dam is also associated with less behavioural and physiological stress. Habituating the foal to separations through frequent, short-term separations prior to weaning appears to have no effect on the subsequent behaviour or physiological responses of either animal and has in some studies been found to intensify the stress response on each subsequent separation.

Hopefully this chapter has provoked you to think about how we can reinvent our managed environments and practices to improve the experiences of the horses involved and give them the best possible chance of a life more ethologically congruent.

References

Bateson, P., & Laland, K. N. (2013). Tinbergen's four questions: An appreciation and an update. *Trends in Ecology & Evolution*, 1–7.

Busby, D., & Rutland, C. (2019). *The Horse a Natural History*. London: Ivy Press.

Cameron, E. Z., Linklater, W. L., Stafford, K. J., & Minot, E. O. (2008). Maternal investment results in better foal condition through increased play behaviour in horses. *Animal Behaviour*, *76*(5), 1511–1518.

Fraser, A. F. (1997). *The Behaviour of the Horse*. Wallingford: CAB International.

Goodwin, D. (2007). Horse behaviour: Evolution, domestication and feralisation. In N. Waran (Ed.), *The Welfare of Horses: Animal Welfare* (pp. 1–18). Dordrecht: Springer.

Greening, L., & McBride, S. (2022). A review of equine sleep: Implications for equine welfare. *Frontiers in Veterinary Science*, *9*, 916737.

Harvey, A. M., Ramp, D., & Mellor, D. J. (2022). Review of the foundational knowledge required for assessing horse welfare. *Animals*, *12*(23), 3385.

Hausberger, M., Henry, S., Larose, C., & Richard-Yris, M. A. (2007). First suckling: A crucial event for mother-young attachment? An experimental study in horses (Equus caballus). *Journal of Comparative Psychology*, *121*(1), 109.

Henry, S., Richard-Yris, M. A., & Hausberger, M. (2006). Influence of various early human–foal interferences on subsequent human–foal relationship. *Developmental Psychobiology: The Journal of the International Society for Developmental Psychobiology*, *48*(8), 712–718.

Henry, S., Richard-Yris, M. A., Tordjman, S., & Hausberger, M. (2009, August). Neonatal handling affects durably bonding and social development: Experimental studies in the domestic foal. *Behavior 2009-31st International Ethological Conference*.

Henry, S., Sigurjónsdóttir, H., Klapper, A., Joubert, J., Montier, G., & Hausberger, M. (2020). Domestic foal weaning: Need for re-thinking breeding practices? *Animals*, *10*(2), 361.

Kentucky Equine Research. (2011). *Lower Critical Temperature for Horses*. Available at https://ker.com/equinews/lower-critical-temperature-horses/

McDonnell, S. M. (2000). Reproductive behavior of stallions and mares: Comparison of free-running and domestic in-hand breeding. *Animal Reproduction Science*, *60–61*, 211–219.

McDonnell, S. M. (2003). *The Equid Ethogram*. Lexington: The Blood Horse.

McDonnell, S. M., & Haviland, J. C. S. (1995). Agonistic ethogram of the equid bachelor band. *Applied Animal Behaviour Science*, *43*, 147–188.

McDonnell, S. M., & Poulin, A. (2002). Equid play ethogram. *Applied Animal Behaviour Science*, *78*, 263–290.

Mills, D. S., & McDonnell, S. M. (Eds.). (2005). *The Domestic Horse: The Origins, Development and Management of Its Behaviour*. Cambridge University Press.

Powell, D. M. (2008). Female-female competition or male mate choice? Patterns of courtship and breeding behavior among feral horses (Equus caballus) on Assateague Island. *Journal of Ethology*, *26*, 137–144.

Ransom, J. I., & Kaczensky, P. (Eds.). (2016). *Wild Equids: Ecology, Management, and Conservation*. JHU Press.

Rossdale, P. (2002). *The Horse from Conception to Maturity*. London: J.A. Allen.

Sigurjónsdóttir, H., van Dierendonck, M., Snorrason, S., & Thórhallsdóttir, A. (2003). Social relationships in a group of horses without a mature stallion. *Behaviour*, *140*(6), 783–804.

Søndergaard, E., & Jago, J. (2010). The effect of early handling of foals on their reaction to handling, humans and novelty, and the foal-mare relationship. *Applied Animal Behaviour Science*, *123*, 93–100.

van Dierendonck, M. C. (2006). *The Importance of Social Relationships in Horses*. Utrecht: Utrecht University.

van Dierendonck, M. C., de Vries, H., Schilder, M. B., Colenbrander, B., Þorhallsdóttir, A. G., & Sigurjónsdóttir, H. (2009). Interventions in social behaviour in a herd of mares and geldings. *Applied Animal Behaviour Science*, *116*(1), 67–73.

Waring, G. H. (2003). *Horse Behavior: The Behavioral Traits and Adaptations of Domestic and Wild Horses, Including Ponies* (pp. viii–292) (2nd ed.) New York: William Andrew Publishing.

Chapter 3

Equine senses and communication

Equine senses

Horses, like most mammals, receive information about their environment through their senses – vision, hearing, olfaction (smell), taste and touch. How this information is organised, interpreted and experienced is known as perception. Although horses and humans share the five most common sensory modalities, sensory evolution in the horse has resulted in distinct visual capabilities, acute sensitivity to touch and highly developed hearing, smell and taste, meaning they are unlikely to perceive their surroundings in the same way we do. Sensory ability, perception and behaviour are closely linked, which is why an understanding of the horse's sensory abilities and how they differ from ours is important not only for the purposes of ethological homeostasis (Fraser, 1983) but also for evaluating welfare and for training and behaviour work. In addition, once we accept that equine sensory abilities are different from those of people, and we recognise that individual equine perception is also affected by factors such as breed, temperament and age, we can appreciate the impact these differences have on each and every interaction between a horse and a human. From human research we know that sensory sensitivity is influenced by the individual differences and relates to their thresholds for noticing, responding to and becoming irritated by stimuli. This also appears to hold true for horses, highlighting the need not only to understand the differing sensory abilities of horses compared to humans but also to differentiate between the sensitivity of each horse and their subsequent requirement for individual care and management (Rørvang *et al.*, 2020).

To date, research on the sensory abilities of horses has mainly focused on vision and hearing. Some studies have been done on olfaction but tend to focus on the influence of smell in reproduction and social behaviour rather than the wider impact of smell on the horse's perception of the world around them. The more we can learn about how horses perceive their environment, their conspecifics and us from a sensory perspective, the better our awareness and recognition will be of what they are likely to find aversive or pleasant, thus enabling more efficient, welfare friendly and ethologically sound training and handling techniques, as well as improving human safety (Hausberger *et al.*, 2008).

Vision

The horse's visual abilities evolved to provide optimal predator detection, and vision is considered the horse's primary used sense. Most predators have their eyes on the front of their head and utilise their narrow, binocular field of vision (they see the same thing with both eyes) to

DOI: 10.1201/9781003581390-4

focus in on their prey. The eyes of prey animals are situated on the sides of the head providing panoramic vision or scanning capacity, giving a visual field of nearly 360 degrees (there's a blind spot directly behind and in front of them). This allows for a large range of vision and greater sight of their surroundings enabling them to keep track of and evade predators. Although their long-range acuity is relatively poor (visual acuity is defined as the ability of the eye to distinguish shapes and details of objects at a given distance), the horse's eye is extremely sensitive to movement. In contrast to the human retina (which is curved), the equine retina is flat and sloped backwards which contributes to peripheral vision and is known as a visual strip. This gives the horse the ability to broadly see the entire horizon but much less below or above it, which may explain why horses often have fears concerning their overhead space as they are limited in what they can easily see above eye level. To bring an object into focus the horse needs to lift, lower or tilt their head in order to bring it in line with the visual strip. When a movement attracts their attention they can either swing the head round to see it with both eyes, tilt the head or move it up or down to place the image on the visual strip. In situations where the horse is unable to or doesn't want to move their head they can roll the eye to get the same effect. A horse has the ability to roll their eye all the way round to look at their hind feet and may show the white of the eye while doing this. This can be misinterpreted as a sign of fear or aggression when actually the horse is simply trying to see what's going on behind them.

Horses have particular difficulty focusing on objects directly in front of them, especially if the object is less than one metre from their head. The horse uses both eyes until an object is within a metre or a little further, when they must then turn or lower the head to continue to observe the object with one eye. The concentration needed to focus closely forward often means they momentarily lose the ability to observe consciously what is going on behind or to the side of them, an important consideration for the safety of handlers. Horses also have the advantage of monocular vision with each eye moving independently and able to see separate things at the same time, which may help to explain a great deal of their behaviour. The lateral position of the eye provides an effective early warning system for the grazing horse by enabling them to see approaching predators from either side of the body. Having seen an object with one eye the horse must make an adjustment in their head, neck or body position in order to bring that object into focus with both eyes so they can assess its level of threat. This sudden awareness can be the cause of many spooks or shies at seemingly small and innocuous objects. The smaller (and closer) the object the more likely it is to suddenly and unexpectedly come into the horse's conscious awareness. Head and neck position are therefore important factors found to affect the visual abilities of horses. Lowering the head brings close objects into sharper view; raising the head allows objects at a greater distance to be seen. However, a horse with a high head carriage loses both side and ground vision, another safety consideration for handlers. When moving, the horse can see even less clearly which is perhaps why young or inexperienced horses often want to stop and look at strange things and can get panicky if they are prevented from doing so, since stopping and orienting to such objects in order to gather information is one of their early lines of defence.

The ability to perceive (make sense of) visual images is dependent upon the amount of information available from the retina, gathered via photoreceptors known as rods and cones. The equine eye is rich in rods, the photoreceptors most sensitive to low light levels, and horses have a higher proportion of retinal rod cells than humans have. They also have a reflective structure called the tapetum lucidum which acts like a mirror reflecting light back into the eye. All of which means they have a far superior capacity for night vision than we have. However, in bright daytime conditions the horse is actually at a disadvantage in comparison to people. The numbers of rods in the horse's retina is significantly greater than the number of cones, the photoreceptors responsible for vision in brighter conditions, and studies have shown that horses see details better on overcast days whereas a human's visual performance is maximised on bright and sunny days.

Horses do have depth perception although it's different to ours because we utilise binocular vision. In essence we see the same thing through both eyes, albeit from a slightly different angle. Humans have a binocular field of vision of around 120 degrees whereas the horse is limited by only having a binocular vision of 55–65 degrees. In other words, the horse can only see the same thing with both eyes in a very limited visual range and that is predominantly below the head. Depth perception studies have shown that horses actually use the separate input from each eye to replicate binocular vision and thus perceive depth and 3D structure allowing them to be able to gauge the distance and height required to clear a jump or obstacle. There is some disagreement about the colour vision abilities of horses. Equine vision is considered dichromatic, meaning the eye contains only two types of cone photopigment instead of three. A lack of the third pigment is generally considered partial colour blindness and it is believed that horses see colours in a similar way to people with red-green colour blindness. Several studies have looked at the colour-seeing abilities of horses in relation to jumping and found that obstacle colour affects how likely the horse is to detect and therefore successfully jump the obstacle. Fences that were all dark or all light proved problematic as did the surrounding colours of the environment; for instance when jumping on grass, blue-coloured obstacles were detected more easily than green ones. Even the angle jumped, the distance jumped and the landing distance have been shown to be influenced by the colour of the obstacle. Research with racehorses demonstrated that white jumps resulted in the largest take-off distance, bright blue jumps resulted in a larger take-off angle and fluorescent yellow fences had shorter landing distances compared to orange ones.

Since our eyes work so differently from our horses it would be safe to assume that our visual perception of the world is also very different. This is a very important consideration for anyone involved in riding, handling or training horses and has a direct impact on both their general learning and behaviour. By appreciating how the horse acquires information from their environment and subsequently reacts to that information we will be better able to identify stimuli that are salient (important) to the horse and therefore attract the most attention (Hall, 2007). The optimal training programme requires that the horse is paying attention to the relevant cues and ignoring irrelevant ones. With knowledge of how the horse's visual system operates we can design and implement training and behaviour modification programmes which match their natural inclinations and present stimuli that are most likely to be noticed and responded to. Visual perception undoubtedly determines the ability to learn visual tasks, and because human visual perception is so different to equine visual perception, consideration must be given to both the visual ability and therefore the behavioural tendencies of the horse in order to improve our understanding of their learning and behaviour.

Hearing

Unsurprisingly there are also marked differences between the way horses and humans hear the world around them. Firstly, they can detect frequencies that we can't – generally softer and higher tones. Secondly, this well-developed sense is aided by having very mobile ears which are capable of moving in a lateral arc of 180 degrees. In this way they can act like funnels, moving independently to collect sounds from different directions and greatly improving aural acuity. Interestingly, studies have shown that large land mammals tend to be good at hearing lower frequencies but horses are the exception. The lowest frequency detectable by the horse is 50 Hz, which is higher than the lowest human detection threshold of 20 Hz. By contrast, at the higher frequencies equine hearing exceeds the highest frequencies that can be heard by people – 33 kHz compared to 20 kHz respectively. This clearly demonstrates that there will be situations where a horse can detect and react to sounds that we are simply unable to hear. One other interesting piece of research found that like humans, horses' hearing ability is affected by age, with older horses showing fewer behavioural reactions to sounds than younger horses.

There is some suggestion that these older horses learn to compensate for any loss in hearing by utilising their other senses more (such as vision) as well as learning the daily routines that allow them to behave to our observation as if their hearing has not changed. But it is worth bearing in mind that there may be some weakening of hearing ability as the horse ages and this can result in voice commands being less detectable.

Given the acute visual and aural capabilities of the horse, maybe it's time to explore our ways of communicating with them in order to determine the most efficient, ethologically congruent and welfare-centric horse-human interactions.

Touch

Horses are extremely sensitive to touch, which is why it is perhaps surprising that very little research has been conducted on their tactile sensitivity, despite it being the main method used in training and handling. The skin is the largest organ in the horse (as well as in humans) and it is therefore the largest of the sensory organs as well as the main communication interface between horses and people. The sensitivity of the skin varies across the body in relation to the distribution of the sensory nerve receptors. The muzzle, neck, withers, coronets, shoulders, lower flank and rear of the pastern are considered the most sensitive parts of the body. At this point it's worth mentioning the vibrissae or whiskers around the muzzle and eyes. These whiskers have different characteristics to hair follicles and are considered sense organs in their own right, meaning that any removal or thinning has the potential to deprive the horse of an important source of sensory information.

Skin is sensitive to both thermal (temperature) and mechanical stimulation (movement). As discussed in the chapter on maintenance behaviours, mutual grooming between two horses is considered a positive behaviour and has been shown to have a calming and social bonding effect. In a similar way, mechanical stimulation of the skin in the form of massage may also be beneficial in alleviating stress. Studies found that a daily massage had a strong positive impact in low to medium stressful situations (vet visits or isolation) than, for instance, playing relaxing music. Research in humans indicates that oxytocin, which is believed to have health-promoting effects, is elevated in the circulation following light touch, light pressure and massage-like stroking. Although research in the horse is limited there is no reason to suggest that it is not also true for them, which is therefore another reason to be mindful when grooming your horse and not just intent on scrubbing off mud, sweat and dirt. In fact, positive tactile stimulation has potential not only as a reward but also as a stress-relieving aid, and there may be unexploited opportunities for using tactile stimuli more than we do currently.

Olfaction

Horses have a highly developed sense of smell, which is often overlooked, both in equine research as well as in applied contexts of training and behaviour therapy. Their noses can move large volumes of air in one breath, trap large numbers of molecules and, because their nostrils are separated and angled separately, they have the ability for stereo olfaction. In response to interesting, strong or usual smells, horses will often exhibit the flehmen response (Figure 3.1), where the neck is extended, the head is raised and the upper lip is curled back displaying the front teeth. This posture facilitates the transfer of inhaled scent molecules into the vomeronasal (or Jacobson's) organ enabling the horse to analyse odours with great accuracy. Depending on the scent detected, various physiological and behavioural reactions may then be triggered, including neuronal signals sent to the amygdala and hypothalamus, which we will learn more about in chapter 6. Such a complex olfactory apparatus suggests that information from odours are a fundamental part of how the horse interprets the world around them. Studies that have examined the role of olfaction in horses to date have mainly focused on reproduction and social

Figure 3.1 Adult horse displaying the flehmen response. Photograph by slowmotiongli.

recognition (King *et al.*, 2022). Interestingly what has been discovered is that despite a common belief that stallions use their sense of smell to determine when a mare is in oestrus, actually they didn't appear to differentiate between odours of mares in oestrus and those who were not. When we consider the ethology of the stallion we realise that detection of odours is not important when belonging to a bachelor group or even when leading a predominantly female harem as only the mares in oestrus will allow the stallion to mount. In this sense the stallions rely on the mare's behavioural response to determine when they are ready for mating rather than on any odour.

Odours are, however, important for social recognition as horses are able to distinguish between individuals based on their smell. Odour profiles of hair samples differ among horse breeds and each horse has been found to have their own odour profile with a certain degree of similarity among family members. Studies of feral and free-roaming horses describe how these animals recognise each other on the basis of body odours as well as odours from urine and faeces. Moreover, horses exhibited more interest in the faeces of horses from whom they had received the highest amount of aggressive behaviours (Krueger and Flauger, 2011). Worth bearing in mind if you are a trainer or behaviourist who is interacting with many horses a day, as they will inevitably leave a scent on clothing and equipment with the potential to cause a reaction in subsequent horses, particularly if a previous horse is prone to behave in an aggressive manner to others. In studies by Sabiniewicz *et al.* (2020) and Jardat *et al.* (2023), horses were presented with samples of human body odour collected under fearful or more positive conditions. The horses showed different behavioural responses to the two types of body odour, which suggests that horses are capable of using their sense of smell to assess the emotional state of people and change their behaviour according to what they detect.

Figure 3.2 Rachaël Draaisma working with a horse doing scentwork. Photograph by Gijs Timmers Photography.

Somebody who is breaking new ground as a practitioner is Rachaël Draaisma (Figure 3.2), who started using scentwork as mental stimulation for horses, something the dog world has done for decades as a form of enrichment, and now runs courses teaching owners and behaviourists how to put this interesting and beneficial form of equine enrichment into practice (Draaisma, 2020).

Taste

Humans breathe through both their nose and their mouth and are therefore able to make connections between smell and taste. Horses on the other hand only breathe through their nostrils (unless they are physically prevented from doing so through injury or abnormality). The tasting organs of the horse are linked to the olfactory system but it is unclear if this also allows them to link smells and tastes. Research has shown that horses are able to detect sweet, sour, salty and bitter tastes, although individual horses vary in their response and preference for each.

Equine communication

The horse is a social species and will actively choose to live together in large herds. As a result, a great deal of their behaviour concerns communication among herd mates, ensuring peaceful, ongoing interaction. Horses are primarily visual communicators and are extremely sensitive to subtle changes in the body language of their companions. The social life of the herd depends on the sending and receiving of signals that other members of the group understand. Some appear to be deliberately sent such as threat faces, others like body postures seem to be an almost involuntary expression of their inner state. A familiar combination of signals that communicates fear is a high head carriage, tense neck muscles, wide eye showing sclera (whites of the eye) and flared nostrils. Body postures and signals like these are very obvious, whilst others are more subtle. In particular, minute and complex communication takes place in the musculature of the face so that tension around the ears, eyes, nostrils, mouth and chin vary according to the context and visually signal a range of both cognitive and emotional states. Recent research is taking note of these more subtle signs leading to the development of tools such as the equine grimace scale and the ridden horse pain ethogram.

Horses also communicate spatially by actively deviating from their current action or route in order to avoid another individual. This has been interpreted by some human observers as submissive behaviour although it may actually indicate individual preferences for certain group members over others or a choice to avoid confrontation, thereby preserving energy and the cohesion of the group. Other signals often interpreted as indicators of submission include licking-chewing and head lowering. At present no published accounts are available of these behaviours existing in the social interactions of feral or free-roaming herds. It has been suggested that these may be displacement activities or comfort-seeking behaviours associated with conflicting motivations in domesticated horses; however, the explanation that best follows the scientific evidence is given by Dr Sue McDonnell who points out that this behaviour is typically observed following a period when a horse has been scared, confused or excited, and thus

Table 3.1 Features and Function of Vocalisations in Horses

Vocalisation	Features	Function
Nicker	Low vocal tone through quivering nostrils, typically between two familiar horses.	Greeting particularly after there has been an absence. Also referred to as anticipatory calls or contact calls.
Whinny	Loud prolonged call of one to three seconds. Begins high pitched and ends low pitched.	Separation calls and to maintain social contact at a distance. Associated with alert and approach from a distance perhaps when contact has been lost with a preferred partner or the rest of the herd. "Where are you?" or "where have you been?"
Squeal	Shorter, higher pitched. Squeals of dominant stallions last longer than subordinates and serve to signal status and fighting ability.	Emitted during moments of excitement or emotional arousal, e.g. when unfamiliar horses meet or during fighting (mock/serious)
Snort	Sudden, strong exhalation from the nostril and may snort several times at perceived threat or novel object.	Associated with olfactory investigation and may be intended to clear the nostrils for better identification of smells.
Roar/scream	High intensity produced during extreme emotional arousal.	Follows earlier more subtle threats that may have been ignored. Can indicate more serious signs of aggression.

in a physiological state of sympathetic arousal. In a training situation, when the trainer takes the pressure off and the horse's physiological arousal starts to abate, their autonomic nervous system (more of this in chapter 6) returns to the parasympathetic, and it is at this point that we often see the licking and chewing response (as salivation returns). Those who suggest that it indicates relaxation aren't wrong, but it's likely that this has been preceded by a stressful experience. Yawning may occur in similar situations.

Studies of free-roaming horses have recorded communication using a body language of subtle signals, feints, expressions, nips and postures. Another, not unsurprising, finding was that vocal language (Table 3.1) was used a lot less than visual communication. Having evolved to live in plain sight on vast, open plains, the ability to remain silent and not draw the attention of predators is crucial. As a result, horses are not a particularly vocal species, but research has identified five distinct horse to horse vocalisations:

Studies on free-roaming Przewalski's horses (Maigrot *et al.*, 2017) found that nickers were most commonly produced in positive contexts such as anticipation of a food reward and affiliative actions, whilst whinnies and squeals occurred more often in negative contexts (agonistic interactions and social separation). Since the type of calls produced differs according to the context (positive/negative), we can draw the conclusion that vocalisations give important information about the emotional state being experienced by the horse in that moment.

Learning how horses communicate with each other is the basis for developing a communication style that horses will understand, and understanding their communication with us, whilst knowledge about equine vocalisations provides insight into their current affective state. An understanding of both of these areas can greatly enhance the horse-human interaction.

Horse-human interactions

Horses read humans in various ways – by our odours, posture, facial expressions, attentiveness and vocal repertoire. It's important to know that they remember previous experiences of interacting with us and this affects their interpretation and reaction to us. While it is difficult to truly know how a horse thinks and feels, paying attention to subtle behavioural signals can give us an insight into how horses view humans and how they prefer us to interact with them.

Studies show that horses choose to approach an attentive person over an inattentive person, and the more positive experiences a horse has with humans, including small actions such as approaching with a happy facial expression, the more positive they are to all future human interactions. Domestication has led to the development of the horse's ability to recognise human emotional expression and by reading these cues they have learned to react accordingly.

Attachment theory

Attachment theory, initially developed by John Bowlby in the context of human relationships, posits that early attachments, especially those between infants and caregivers, are crucial to the individual's development and affect their future social relationships. Bowlby suggested that children are biologically predisposed to develop attachments with caregivers as a means of survival. This foundation was empirically explored by psychologist Mary Ainsworth, who attempted to elaborate on Bowlby's ideas with her "Strange Situation" experiment. This structured observation demonstrated how infant attachment security is expressed through children's behaviour when reunited with their mother after a brief separation, highlighting secure, avoidant, and ambivalent attachment styles. Lundberg *et al.* (2020) explored this in the context of the horse-human relationship, using a modified version of the "Strange Situation", and found evidence of attachment-type behaviours towards humans, more so in horses trained primarily with positive reinforcement.

Parallel to Bowlby's work on humans, Harry Harlow's experiments with rhesus monkeys provided further insight into the importance of attachment. Harlow showed that infant monkeys formed attachments based on physical and emotional comfort, choosing to seek out cloth surrogate mothers over wire surrogates that provided food but no comfort, thus underscoring the importance of warmth and security in attachment development.

We can extrapolate the principles of attachment theory to equine behaviour, since it's evident that horses, as social herd animals, also exhibit attachment behaviours. In the wild, foals show a strong attachment to their mothers, which ensures not only immediate protection and care but also social learning and integration into the herd. These early attachments influence their future social dynamics, aligned to the foundational theories observed in humans and other primates. In domestic settings, horses form significant attachments not only with other horses but also with humans, and these relationships can mirror the secure base seen in the mother-child and monkey-cloth mother experiments. Horses clearly seek comfort and security from their human caregivers, who we might say effectively become surrogate herd members. Secure attachment in horses can be observed through behaviours such as greeting, following, and showing a preference for specific individuals, whether equine or human.

These attachments have a significant effect on training and handling, since horses that are securely attached to both equine conspecifics and significant human friends (Figure 3.3) tend to show higher cooperation and reduced stress responses, similar to securely attached children who explore their environments with greater confidence. Understanding these attachment dynamics is crucial for ethological and relational approaches to horse training (Hartmann *et al.*, 2021) and behaviour therapy, emphasising the need for trainers and caregivers to provide security and build trust in order to foster optimal interspecies relationships and wellbeing.

Figure 3.3 Connection and communication between Drago and Melody. Photograph by Maia Nootens, Equine Photographer.

Horse-human communication

We're well aware that horses do not use verbal language but we must also accept that humans cannot speak "horse". This means that working together the horse and human must create a system of communication using a medium they can both understand, in essence a third language (Brandt, 2004). Because horses rely on their bodies to transmit and receive information they are highly skilled at reading (and using) body language, so this is a helpful place to start.

Communication is developed from consistent, ongoing interactions designed to have a positive effect both in the moment and over time. Knowing that our interactions with the horse are cumulative and keeping our relations positive will result in the horse maintaining a better outlook on people in general (Birke and Thompson, 2018). Our body posture, attentiveness and transmission of communicative cues are all key elements in creating and maintaining this positive relationship. Understanding the horse's ethogram, their sensory capabilities, how they perceive the world and how they interpret people is fundamental to improving the relationship between horse and human and the wellbeing of both (Merkies and Franzin, 2021). Working towards optimising this relationship is essential to improving the welfare, training, husbandry, management and experiences of horses living in the human world (Argent and Vaught, 2022).

References

Argent, G., & Vaught, J. (Eds.). (2022). *The Relational Horse: How Frameworks of Communication, Care, Politics and Power Reveal and Conceal Equine Selves* (Vol. 24). Brill.

Birke, L., & Thompson, K. (2018). *(Un)Stable Relations Horse, Humans and Social Agency.* Abingdon: Routledge.

Brandt, K. (2004). A language of their own: An interactionist approach to human-horse communication. *Society & Animals*, 12–14.

Draaisma, R. (2020). *Scentwork for Horses*. CRC Press.

Fraser, A. F. (1983). Processes of ethological homeostasis. *Applied Animal Ethology, 11*(2), 101–110.

Hall, C. (2007). The impact of visual perception on equine learning. *Behavioural Processes*. doi: 10.1016/j.beproc.2006.09.017

Hartmann, E., Rehn, T., Christensen, J. W., Nielsen, P. P., & McGreevy, P. (2021). From the horse's perspective: Investigating attachment behaviour and the effect of training method on fear reactions and ease of handling—a pilot study. *Animals*, *11*(2), 457.

Hausberger, M., Roche, H., Henry, S., & Visser, E. K. (2008). A review of the human–horse relationship. *Applied Animal Behaviour Science*, *109*(1), 1–24.

Jardat, P., Destrez, A., Damon, F., Menard--Peroy, Z., Parias, C., Barrière, P., Keller, M., Calandrau, L. & Lansade, L. (2023). Horses discriminate human body odors between fear and joy contexts in a habituation-discrimination protocol. *Scientific Reports*, *13*(1), 3285.

King, S. R., Schoenecker, K. A., & Cole, M. J. (2022). Effect of adult male sterilization on the behavior and social associations of a feral polygynous ungulate: The horse. *Applied Animal Behaviour Science*, *249*, 105598.

Krueger, K., & Flauger, B. (2011). Olfactory recognition of individual competitors by means of faeces in horse (Equus caballus). *Animal Cognition*. doi: 10.1007/s10071-010-0358-1

Lundberg, P., Hartmann, E., & Roth, L. S. (2020). Does training style affect the human-horse relationship? Asking the horse in a separation–reunion experiment with the owner and a stranger. *Applied Animal Behaviour Science*, *233*, 105144.

Maigrot, A. L., Hillmann, E., Anne, C., & Briefer, E. F. (2017). Vocal expression of emotional valence in Przewalski's horses (Equus przewalskii). *Scientific Reports*, *7*(1), 8779.

Merkies, K., & Franzin, O. (2021). Enhanced understanding of horse-human interactions to optimize welfare. *Animals*, *11*.

Rørvang, M. V., Nielsen, B. L., & McLean, A. N. (2020). Sensory abilities of horses and their importance for equitation science. *Frontiers in Veterinary Science*, *7*, 633.

Sabiniewicz, A., Tarnowska, K., Świątek, R., Sorokowski, P., & Laska, M. (2020). Olfactory-based interspecific recognition of human emotions: Horses (Equus ferus caballus) can recognize fear and happiness body odour from humans (Homo sapiens). *Applied Animal Behaviour Science*, *230*, 105072.

Part 2

How horses learn

Chapter 4

Fundamentals of learning theory

Learning theory and concepts

Why is it important to understand the theory of how horses learn? Because, in most of our interactions with horses we're usually trying to get them to do things that the horse wouldn't naturally prefer to do, or be motivated to do, at the time that we want them to do it. If we want to enjoy our horses and help other people enjoy their horses, it's something that's useful to know about.

At this point it's worth clarifying what we mean by learning and how we know that the horse has learned. Learning involves some sort of change in behaviour that lasts. When we are able to observe a consistent and repeatable change in behaviour then we know learning has taken place. Learning theory attempts to explain how such changes occur and what has to happen in order for behaviour to change. Motivation is one key factor that influences whether a learned behaviour will be performed; another is the learning contingencies. A learning contingency is a dependent relationship between events which in simple terms means "depends upon". For example, the horse getting a treat depends upon them standing still at the mounting block. More on both of these topics later in the chapter.

Other common terminology that you may come across while studying learning theory:

- Conditioning – another way of talking about learning and used interchangeably.
- Response – any action performed that can be observed and measured.
- Stimulus – anything in the horse's environment that could cause them to exhibit or display some type of behaviour. It could be a person, an object (food bucket), something visual (a saddle), a noise (feed room door latch), a smell (another horse's droppings), or a touch (our hand on their hip), or it could be a situation or an event such as arriving at a show or going through the gate to leave the yard on a ride.
- Stimulus collage – the horse's environment is filled with sensory input from sight, hearing, touch, taste and smell as well as bodily conditions such as hunger, thirst, temperature – collectively known as the stimulus collage.
- Consequence – an action or event that occurs after a response which can affect how often that behaviour occurs in the future.

DOI: 10.1201/9781003581390-6

- Appetitive – anything that the horse considers a good thing, for example feels good, tastes good.
- Aversive – anything that the horse considers a bad thing, anything unpleasant, painful, annoying, uncomfortable.
- Discriminative stimulus (cue) – a consistent signal of some kind that helps a horse to reliably predict when and where a specific behaviour is likely to be reinforced, or indeed punished. It predicts a consequence if a specific behaviour is performed. An example of this is the application of the leg as a cue for the horse to move forward. The horse learns that when they feel pressure (to an uncomfortable degree) of the rider's leg and they move forward, the pressure will be released. Thus the leg pressure is the signal that tells the horse, "if you move forward now, this pressure will be removed" (negative reinforcement).
- Stimulus control – continuing the prior example, when the horse has learned to move forward off the leg, and does not move off they do not perceive that pressure, that behaviour is said to be under stimulus control.

Non-associative learning

Non-associative learning involves exposing the animal to a single stimulus to which it can either habituate or become sensitised.

Habituation: nothing to see here

Horses' learning is driven by the need to stay healthy, survive and reproduce, all of which are biologically significant. In a world where resources may be scarce there is also a drive to conserve energy by learning what to approach and what to avoid, when to gallop away from a threat or when to ignore something that has proven to be of no consequence. The technical term for the process of learning to ignore the stimuli of no consequence is habituation or "getting used to things".

Habituation is the simplest form of learning and means that the horse avoids wasting energy on insignificant events but remains alert to changes in the environment that may represent a threat. In evolutionary terms habituation is important for survival: horses must learn quickly what to pay attention to, and what is safe to ignore, in order to conserve energy for survival (Busby and Rutland, 2019).

A stimulus – something that can be noticed by the horse, be it an object, person, animal, sound or smell – comes to the horse's attention. If no intrinsic fear is experienced, or if the horse can remain, or move, far enough away, then, crucially, over time and with repeated presentations, the horse will be able to get closer and closer, "get used to", to the stimulus. Horses' ability to habituate has been helpful for humans throughout the duration of our relationship with them for all the potentially threatening activities and situations that we ask them to engage in, whether going into battle, pulling a hay cart, hacking out in traffic or loading into a trailer.

Habituation is context-specific (Reid, 1996), so a horse might habituate to a stimulus in one specific context or environment but still react to the same stimulus in a different context. For example a horse who has habituated to a noisy train passing by their own familiar field may still react fearfully to that stimulus while being ridden along the road next to a railway line. The context-specific nature of habituation is important because it shows that the learning associated with habituation isn't just about the stimulus itself but also involves the environment and conditions in which the stimulus is experienced (Figure 4.1). This specificity can be crucial for free-roaming equids, where the safety or threat level of a stimulus might vary depending on the context.

Figure 4.1 Habituation is context-specific; this horse has an unfamiliar experience and reacts fearfully. Photograph by Paul Briden.

In situations where a stimulus initially elicits a fear response, a horse may habituate to it over time if repeated exposures demonstrate that the stimulus poses no threat. However, the core concept of habituation is that a response reduces due to repeated exposure, regardless of the nature of the initial reaction, whether fearful, curious or indifferent (Figure 4.2).

Habituation is subject to spontaneous recovery, where, without some degree of continuing exposure, a response that had habituated may recur.

Figure 4.2 Habituation: an unrestrained horse uses their mouth to explore a novel object. Photograph by Fotema.

Sensitisation: something to worry about

In contrast to habituation, we may see its opposite, where the response to a stimulus increases after repeated presentations, where that stimulus is intrinsically unpleasant or aversive. No association needs to be made with another stimulus; sensitisation simply enhances the response to a stimulus due to repetition or intensity of exposure, without the need for pairing or consequence.

Sensitisation is evolutionarily adaptive because it ensures the horse notices and reacts to something that is potentially threatening to their health, survival or ability to reproduce.

A horse is sensitised if they become increasingly reactive to an object, an event or an experience. In essence, things that weren't worrying become worrying: a stimulus comes to the horse's attention. If the horse experiences fear, and if the stimulus is encountered repeatedly, then sensitisation occurs and you'll see the fear response repeated and likely to increase. In the earlier train example, if the horses were turned out in a small paddock where even the furthest point from the railway line was not far enough away for them to feel safe, then the fear response would still be elicited and they would be likely become sensitised to the sight, sound and smell of trains. In this case their fear response would increase rather than decrease (Busby and Rutland, 2019).

Associative learning

Associative learning is the process that takes place when a horse learns that their response is associated or paired with a stimulus (classical conditioning) or that their behavioural response is associated with a consequence, whether good or bad (operant conditioning).

No discussion of this area of psychology is complete without an introduction to the men who contributed its foundational theories: Edward Thorndike, B. F. Skinner and John B. Watson were early 20th-century researchers who shaped our understanding of human and animal behaviour. Edward Thorndike is best known for the development of the law of effect. This principle posits that behaviours followed by satisfying consequences tend to be repeated, while those followed by unpleasant consequences are less likely to be repeated. Thorndike's emphasis on the importance of consequences laid the groundwork for later behaviourist theories.

B. F. Skinner expanded on Thorndike's work, developing the theory of operant conditioning and demonstrating how behaviours could be modified by systematically managing rewards and punishments. John B. Watson theorised that all behaviours are acquired through conditioning, and his famous experiment, the Little Albert study, demonstrated that this extended to emotional responses, something behaviourists are aware of every day in their clinical practice.

Ivan Pavlov was a Russian physiologist best known for his discovery of classical conditioning during his research on the digestive system of dogs. He observed that dogs would salivate not only when food was presented but also in response to stimuli associated with food, such as the footsteps of an approaching technician.

Pavlov's experiments, where he systematically paired the sound of a metronome or a bell with the presentation of food, showed that animals could learn to associate a neutral stimulus with a significant one, leading to a learned response (salivation in this case) to the previously neutral stimulus (Pearce, 2008).

Classical conditioning

Classical, Respondent or Pavlovian conditioning is what happens when a horse learns associations between stimuli and that those associations are significant and predictable (Baldwin and Baldwin, 2000). Consider how some horses react to a car driving onto the yard in the morning, by banging on the stable door in anticipation of receiving breakfast. The neutral stimulus of the car arriving at the yard has been paired so often with the horse receiving their breakfast that the car has become a conditioned stimulus (CS) resulting in the banging of the stable door (conditioned response or CR). Classical conditioning uses the process where pairings are subconsciously associated between neutral stimuli in the environment and the arrival of a biologically important stimulus. If a neutral stimulus (the car) reliably predicts the biologically important stimulus (food), then the horse will soon begin to make a similar response to the neutral stimulus as it does to the biologically important one.

The following are examples to help familiarise you with the terminology used in classical conditioning and how it works in practice.

Example 1

An unconditioned stimulus (US), something that happens in the environment, such as a big, loud, noisy vehicle going past, elicits an unconditioned response (UR); the horse is startled, fearful and runs away. The vehicle happens to be a tractor and at the first encounter it is a neutral stimulus (NS). However, the tractor drives past the horse's paddock every day making lots of noise. The horses experience fear each time it drives past, and they cannot run away far enough to feel safe. The tractor becomes a conditioned stimulus for "run away" (or attempt to – often horses are prevented from "escaping" to safety by some form of restraint, be it fencing or equipment). Now the scary tractor has become a conditioned stimulus (CS) eliciting the conditioned response (CR) of running away, or at least attempting to. We also describe this as the horse having developed a conditioned emotional response – fear of tractors.

The tractor has now taken on the qualities of something scary. This may seem obvious to us but it's important to know that objects in themselves aren't scary; it is the actions or experiences that are associated with them that make them so. If a naïve horse saw a tractor for the first time and nothing scary happened, and they could investigate at their own pace (which usually means being able to approach and move away), then rather than developing a conditioned fear response they would instead habituate to the tractor.

By repeated pairings of two stimuli/events, in this example the unconditioned stimulus (US) of the tractor with the horse's natural response (UR) to run away from the big, noisy thing, the tractor has become a conditioned stimulus for a conditioned response of fear-based behaviour (stopping, spinning, running away).

Example 2

A horse is being ridden up to a stream (NS). The horse hesitates in order to investigate, and the rider kicks the horse on (US) and the horse feels pain/discomfort (UR). The horse has now made an association with the stream, the rider's actions (kicking) and their experience (pain/discomfort). Being ridden up to the stream is now a conditioned stimulus (CS) and elicits a conditioned response (CR) of behaviour typically described as napping (stopping, spinning, resisting the rider's attempts to send them forward). The stream (was NS) has become a CS and has taken on the qualities of a painful stimulus associated with discomfort from the rider kicking, and from the conflict of trying to move off the leg aid but not wanting to get closer to the stream. In this way napping behaviour becomes automatic as soon as the horse feels the rider increase the leg, in this context of being asked to get closer to the stream. We might not think that a bit of leg to go into a stream is a big deal but the proof is in the behaviour. If the napping behaviour increases, then we can conclude that the horse has had a bad experience via the conflict created by being made to approach the stream too quickly without being given sufficient time to investigate it.

Example 3

The horse has an innate fear response to a hissing sound (US) – it might be a dangerous snake! Their natural response upon hearing a hissing sound is to startle and run away (UR). The horse's owner, in an attempt to alleviate the nuisance of flies, buys a spray bottle of fly repellent (NS). The first time the spray bottle is used to cover the horse in fly repellent it makes a hissing noise (US). This elicits an intrinsic startle and run away response in the horse (UR). After several attempts to spray the fly repellent onto the horse, whilst also restraining them with a headcollar and lead rope, the fly spray bottle becomes a conditioned stimulus (CS) for what is a conditioned response (CR) of "attempt to run away" (escape/avoid). Now, even before any spraying has taken place, the horse tries to avoid the bottle as soon as the owner holds it up to spray the repellent.

Horses commonly learn to make fearful associations with objects or experiences (McGreevy and McLean, 2010). Perhaps the horse has had a bad experience being clipped or being loaded or being examined by a vet. We often restrain horses to carry out these procedures so that their escape/avoidance option of "move away to a safe distance as decided by them" isn't available, therefore these sometimes unpleasant procedures are unavoidable (Doherty *et al.*, 2017). Careful behaviour therapy can reverse this response using a process termed "systematic desensitisation and counterconditioning". This involves deliberately pairing the stimulus (e.g. the trailer) with positive experiences, such as food, at the same time as working through a process of gradual exposure to the loading experience, until the horse starts to experience positive emotions in a trailer-loading context. We'll talk more about desensitisation and counterconditioning in chapter 10.

Operant conditioning

Operant conditioning, also known as instrumental learning, or learning by trial and error, describes the associations formed based on the consequences the horse experiences when responding to the stimulus (Reid, 1996). In other words, the horse learns that their behaviour has consequences. With operant conditioning, what the horse does is critical to what happens next. We can use this process to help the horse learn to do something differently using the principle of reinforcement: "you do this" (behaviour/response) "and then that happens" (consequence). Reinforcement increases the likelihood of the behaviour being performed in future, and it is based on whether the consequence is pleasant (appetitive) or uncomfortable (aversive). Once the horse has learned that something they do results in an appetitive consequence for them, they will be more likely to repeat it. They will also learn to do something if it results in being able to avoid an aversive consequence.

The important point here is that the effect of the reinforcer depends on how it is perceived by the horse, not how it is intended by the handler (Mills, 1998). If the two match up, that's great, but this is one of the places where a training plan can come unstuck, and it's the horse's perception that matters. If a trainer thinks they're using an appetitive stimulus (food) to train a horse to lead, but the horse is paying more attention to avoiding the aggressive horse following them from the other side of the fence, then the horse's leading behaviour is actually being shaped by the motivation to avoid something unpleasant: proximity to the unfriendly horse. This also introduces the idea of salience, which we will look at in more detail in the next chapter. If a response made to a stimulus results in a pleasant experience, then the horse learns the value of that response and will be more likely to make the same response (as opposed to all the others it could make) next time they encounter the same stimulus. This is referred to as positive reinforcement. The positive association between that particular stimulus and that particular response is what reinforces (strengthens) the behaviour. In essence, the addition of a reward for that behaviour means it is likely to occur more often in that situation in the future.

In negative reinforcement an unpleasant or uncomfortable stimulus is applied until the horse responds with the desired behaviour, and then that unpleasant stimulus is removed. This involves the horse learning to escape from or avoid the discomfort. It's the principle behind teaching the horse to move away from pressure, such as when you use a leg aid to ask the horse to go forward. The aid (or cue) in the form of pressure from the leg is removed as soon as the horse takes a step forward. The removal of the unpleasant stimulus (pressure from the leg) results in the horse being more likely to make that response (step forward) the next time they meet the same stimulus. In essence, the removal of the discomfort reinforces (strengthens) the behaviour due to the association between that particular stimulus and that particular response. You could think of it as the horse being "prodded" into action to escape the discomfort.

If the horse's behavioural response to a stimulus results in an unpleasant or uncomfortable experience, then the horse learns the undesirability of that response and will be less likely to make that particular response next time they meet the same stimulus. This is referred to as positive punishment. The unpleasant consequence (e.g. a reprimand) is what influences the horse's behaviour. In essence the addition of a "punishment" for that behaviour in that context, over time, means it is less likely to occur. The horse has learned to avoid behaviour that is punished.

Negative punishment is when the horse's response to a stimulus results in the removal or withdrawal of something that they value. This is how they learn the undesirability of that response and will therefore be less likely to make that particular response next time they meet the same stimulus. For instance, if a horse is constantly mugging you for attention then the removal of something valuable to the horse could include you walking away or ceasing contact. This weakens the mugging behaviour (the horse learns to do it less, or not do it at all) because the consequence of the mugging is that the thing the horse values is removed.

In summary, both *positive* reinforcement and *positive* punishment involve the *addition* of something to the interaction, be it a reward or a reprimand/aversive stimulus. Whereas, *negative* reinforcement and *negative* punishment require the *removal* of something from the interaction, be it pressure or something the horse values. The terms "positive" and "negative" here are used in their quantitative scientific sense where "positive" means "to add" and "negative" means to take away (Table 4.1). You've already seen that we also talk about meanings of positive and negative in a qualitative sense when we talk about positive and negative emotions, so it is helpful to get these two constructs clear in your mind at an early stage of learning about learning theory!

Using operant conditioning to train horses is not without risk; not least because it is how the horse perceives the reward or punishment rather than the intention of the trainer that matters, but also because unpleasant stimuli are involved in both negative punishment and positive punishment. Consider the horse who constantly kicks the stable door. The owner may resort to shouting at the horse for banging, and the horse stops. However, have they stopped because of the verbal reprimand (punishment) or have they stopped because they got the attention they were seeking (reward/positive reinforcement) when they first started banging? The answer really lies in the long-term change in behaviour: did the horse's door-kicking behaviour reduce over time? It's likely that the shouting had a positive punishment effect. Did the behaviour increase? Possible that owner attention had a positive reinforcement effect.

In addition, there is the risk of emotional frustration because appetitive stimuli (rewards) are used in both positive reinforcement (given – something added) and negative punishment (something removed or withheld). If the timing of food delivery is poor there is a risk that the horse is left wondering what they have to do to get the reward from you! If used as a punishment (e.g. withholding food) the horse can become frustrated because they haven't evolved to wait for food, usually they just put their head down and there it is! And if we are giving a horse food for doing what we want them to do, there is an obvious corollary that that food is withheld until "the right thing" is done. Our horse knows that it's there, they're trying to work out what it is they need to do to obtain it, and it's easy for frustration to build up in these situations.

Equally, negative reinforcement often has some degree of mental discomfort as the driving emotion for the behavioural response. It takes a skilled and sympathetic trainer to effectively use negative reinforcement so that any discomfort is kept to the minimum required to train the desired response and doesn't cross the line into fear/pain/frustration/confusion. The other training complication to be aware of with operant conditioning is that when a horse is at the stage of having learned that something they do will gain a reward, but they don't know for sure which of a selection of behaviours that something is, they may perform a different response which can be interpreted as wilful disobedience or "playing up" by the unskilled trainer and subsequently punished. In reality, it is simply the horse trying to find the "right" response that from their side of the training relationship gets them the reward they know is available. Our role as trainers is to ensure that we set up the training environment for our horse to succeed, and to understand how to apply learning theory concepts in ways that avoid frustration and confusion in our horses.

Positive reinforcement involves the presentation of a pleasant consequence when the response is performed. This serves to increase the likelihood of that response in the future.

Negative reinforcement involves the removal of an unpleasant consequence when the response if performed. This also serves to increase the likelihood of that response in the future.

Positive punishment involves the presentation of an unpleasant consequence when the response is performed. This serves to decrease the likelihood of that response in the future.

Negative punishment involves the removal of a pleasant consequence when the response is performed. This also serves to decrease the likelihood of that response in the future.

Table 4.1 Quadrants of Operant Conditioning

Positive reinforcement	Positive punishment	Negative reinforcement	Negative punishment
Add something pleasant (reward = good consequence) Behaviour is more likely in the future e.g. giving a food reward, or scratch, or a feel good stroke	Add something unpleasant (reprimand = bad consequence) Behaviour is less likely in the future e.g. flicking the lead rope to stop a horse barging	Remove something bad (removal = good consequence) Behaviour is more likely in the future e.g. removing uncomfortable pressure of a spur once the horse has moved forward	Remove something good (removal = bad consequence) Behaviour is less likely in the future e.g. withholding a food reward until mugging has stopped

Something that it is useful to be aware of when training with the operant quadrants is the idea of a discriminative stimulus (or cue) (Baldwin and Baldwin, 2000). A cue helps the horse to reliably predict when and where a specific behaviour is likely to be reinforced. A whip is a classic example of this. Before I learned about behaviour I always noticed that my horse walked out "better" (as I thought) when I carried a whip, compared with if I didn't carry one. Actually, he'd learned that the whip was a reliable predictor of negative reinforcement.

Later on we will learn that a conditioned stimulus (CS), which you read about in the classical conditioning section, can also be a cue to elicit behaviour that we would like our horse to express.

The Premack principle

The psychologist David Premack took operant conditioning and the principle of reinforcement a step further by exploring how one behaviour could reinforce another. Premack observed that an individual naturally engages *more* frequently in behaviours that are more rewarding, and *less* frequently in behaviours that are less rewarding. In other words, the more rewarding the behaviour, the higher the likelihood the individual will engage in them. The Premack principle is a theory of reinforcement that states that a less desired behaviour can be reinforced by the opportunity to engage in a more desired behaviour (Herrod *et al.*, 2023). For example, a horse is schooled in the arena for ten minutes (less desired) in order to get to eat grass or be untacked and allowed to roll (more desired) when finished.

In essence, the Premack principle is based on the idea that when a horse is able to engage in a wide variety of activities they will have a natural, and indeed individual, tendency to allocate more time to some than others. For instance, one horse when turned out in the paddock may choose to engage in eating first before greeting or interacting with their other field buddies, while another may prefer to do their socialising first and then eat. By understanding our horse's choices/preferences we may be able to predict when the opportunity to engage in one activity will serve as a reinforcer for another activity. It's the same principle used by parents worldwide – "eat all your greens and you can go back to your screens". In essence, a good thing for young people (screen time) can reinforce engaging in (for them) not such a good thing (eating your vegetables)! However, the caveat is that a horse's motivational state makes a significant difference as to whether an activity will be reinforcing. A horse more interested in staying with their friends than eating will be unlikely to find going away from the herd into a trailer in order to get a bucket of nuts very motivating!

Natural Horsemanship

Natural Horsemanship started to gain popularity in the 1990s as a purported alternative to some types of traditional training techniques which were often criticised as being unnecessarily harsh. The premise was that Natural Horsemanship trained horses based on their natural instincts and ways of communicating within a herd. In essence these techniques are rooted in the application of pressure which is released when the horse performs the desired behaviour, which, as you'll now know from reading the foregoing sections in this chapter, fall within the negative reinforcement quadrant.

This is clearly demonstrated by the use of a round pen. The horse has pressure applied, by being chased or driven by the trainer, to encourage the horse to walk, trot or canter round until they show what were previously considered to be signs of "submission" (licking, chewing) but are now understood as indicators that the horse has experienced stress. The pressure is released by the trainer becoming stationary or turning away. The horse learns to avoid the unpleasant stimulus of being "driven" round the pen by working out which behaviour(s) they need to perform for the pressure to be released. As always, timing is critical because the horse learns to perform the behaviour required for the *release* of pressure rather than learning by its continued application.

There has been some criticism of Natural Horsemanship as simply a rehashing of traditional methods like lunging. Others feel it uses techniques that have the potential to frighten the horse. From a scientific standpoint the effects of this set of training techniques are based on the application of negative reinforcement in just the same way as traditional methods use the application and release of pressure from a leg, whip or bit. It's also worth pointing out here that the other side of the coin from negative reinforcement is positive punishment: if a horse is being trained using aversive stimuli, that is tools or situations that are capable of causing discomfort, pain or distress, be they leg pressure, whip, bit or the experience of being driven around an inescapable space, then it is just as possible for trainers to use these stimuli to punish behaviour, whether intentionally or inadvertently.

Clicker training

Clicker training is another method gaining in popularity, and interestingly it makes use of both classical and operant conditioning. Classical conditioning is used to create a link, or a pairing, between the clicker (typically a small metal box; some trainers use a tongue "cluck") and a reward (usually food). A clear and precise "click" is sounded just before the horse receives a reward (a food treat or body scratch). This creates a positive association between the environmental marker (the click) and the pleasant experience/reward (a food treat or body scratch), and the clicker takes on the qualities of the primary reinforcer and biologically relevant stimulus (food/pleasant touch). Once the horse associates the click with the reward, this can be used to "mark" the behaviour that is being trained and the horse can be rewarded for correct responses. This training method has its roots in the positive reinforcement of a behaviour, once that behaviour has been performed. Clicker trainers typically ignore incorrect responses and focus solely on rewarding correct responses; however, importantly they will also set up the learning environment to make the desired behaviour easy for the horse to perform and unwanted behaviour less so.

One of the main benefits of clicker training, and the reason it can be so effective as a training tool, is that it uses a rare sound that the horse is unlikely to encounter elsewhere. The horse only ever hears the clicker when the trainer uses it to mark the desired behaviour. The "click plus primary reinforcer" is contingent on the horse giving the correct response to the stimulus and therefore has a strong effect on learning. However, this method has drawn criticism from those who believe the risk of food-based reward training is that it creates horses who become over-aroused or aggressive in expectation of, usually, a biologically relevant substance (food).

Others focus this limitation on the argument that horses have not evolved to work things out in order to obtain a food reward and hence this type of training can lead to frustration.

Factors that affect learning

Schedules of reinforcement

A reinforcement schedule describes the frequency with which the reinforcement is applied (Reid, 1996). Initially when the association is being established the horse needs to be rewarded every time they perform the desired response – this is referred to as a continuous schedule of reinforcement. Once the association is established, psychologists have found that a variable reinforcement schedule is more effective. The reinforcement is given less frequently and the horse is not rewarded every time they perform the behaviour. When the horse performs a behaviour that has previously been rewarded, an absence of the reward causes them to try again, as though checking. Whilst continuous reinforcement schedules are effective at eliciting a behaviour from the horse, sometimes the performance of that behaviour diminishes as the horse offers the bare minimum in order to obtain the reward. A variable reinforcement schedule motivates the horse to perform the behavioural response every time a particular signal (cue) is given, in the same way that somebody playing a game of chance keeps on trying because they don't know when it will be their turn to win. The unpredictability of the variable schedule can also prompt the horse to make more effort in their response.

Successive approximations/shaping

Once the desired behavioural response has been linked with the reward, the association can be further refined by using successive approximations or shaping. Shaping is the process by which a new behavioural response is established through a series of stages, or successive approximations, towards an end goal (Baldwin and Baldwin, 2000). The process of shaping involves the following steps:

1. Break down the desired manoeuvre into a number of behavioural responses.
2. Work on one response at a time, ensuring that each step is within the capabilities of the horse and can be performed within the limits of the situation and time available.
3. When moving on to a new element, relax the standards on the previous ones.
4. If at any time the horse appears to be having problems with the task, take them back to a previous stage of training.

Behaviourists often use this type of step by step approach when reintroducing a horse to something to which they had previously formed a negative conditioned emotional response, starting with a small and easy version of what they want the horse to be able to do, and building up one new element at a time until the horse is comfortable with the whole "picture".

Some trainers use a version of shaping known as "free shaping", where the horse is given free rein (pardon the pun!) to try out any behaviours they like, and they are marked (clicked) and rewarded whenever they "get it right". A bit like a horse-human version of the children's game of "hot or cold"!

Emotional aspects

How our horse is feeling is a crucial element of training. We must pay attention to the horse's emotional state before anything else. Negative emotions can present barriers to learning, including emotional states such as confusion, overexcitement, irritation or distraction (when the horse's attention is elsewhere). These all relate to a horse in a state of physiological over-arousal; that is, they are neither calm nor relaxed and therefore not in an optimal mental state for learning to take place (McGreevy and McLean, 2010).

The two aspects of motivation

Understanding the interplay between appetitive (goal-seeking) and consummatory (goal-achieving) behaviours, alongside the role of feedback mechanisms, is essential for comprehending not only how horses manage their interactions with the environment but also how and why they decide whether to engage with us in our training activities, or with what degree of enthusiasm to engage.

Evolutionarily, horses need to decide how to spend their time meeting their basic needs, such that their activity affords them that all important opportunity to survive and reproduce. Motivation helps a horse to make a decision about what activity to engage in next, and these activities have two phases: appetitive (goal-seeking) and consummatory (goal-achieving).

The appetitive phase is characterised by the horse's search for resources such as shade, water or food. For example, consider a horse that feels uncomfortably hot; they enter an appetitive phase where the motivation to find shade increases. This is driven by the phenomenon of positive feedback – the discomfort of heat intensifies their efforts to locate a cool area, thereby increasing their level of activity and urgency as they search. Once they find a suitable tree providing shade (Figure 4.3), they transition into the consummatory phase (Mills and Nankervis, 1999). Here, they achieve their goal and the immediate discomfort of the heat diminishes. Engaging in this behaviour triggers negative feedback, where the motivation to continue searching for shade decreases since the need is already met.

Figure 4.3 Consummatory behaviour: appreciating the shade, goal achieved. Photograph by E A Given.

Feedback mechanisms are critical in regulating how horses interact with their environment. Positive feedback in the appetitive phase serves to escalate the behaviour that leads to acquisition of a resource, enhancing survival strategies by pushing the horse to fulfil essential physiological needs. Conversely, negative feedback during the consummatory phase helps conserve energy by reducing or stopping the activities once the need is satisfied, ensuring that the horse does not expend unnecessary energy once comfort is restored.

Understanding how these motivation phases and feedback loops influence behaviour is important for horse owners and behaviourists in two ways: it provides insight into their natural regulatory systems, and it helps us to evaluate our training activities in terms of how we are attempting to motivate a horse to engage with us. When we ask ourselves, "why did my horse do that, at that particular time?" then, as well as explanations based in learning theory or fear/pain-related avoidance, we need to consider their own agentic and survival-based motivations too. By appreciating these mechanisms, we can better predict and manage their behaviour in their varied domestic settings, whether in a training session or during turnout, to ensure good welfare and efficient management.

Predictability

Horses always need to know "what happens next" in order to be able to learn effectively. They need to know – is what happens next going to be good? If so, "I want to stay round and do what's needed to obtain the good thing". If, however, what happens next is going to be bad, "what do I need to do to avoid or escape it?" One very simple example of predictability that I have used with every horse I've worked with is to always give a carrot when turning out, after taking off their headcollar. This avoids the risk of a fresh horse just turning and whizzing off up the field, with the possibility of a buck and heels flying in my direction!

Making learning easy

One of the most famous sayings in animal training is "Pavlov is always on your shoulder" (with B. F. Skinner, "the father of operant conditioning", on the other!), which means that even when we think we're focusing on operant conditioning (training by consequences), there is always a classically conditioned association, or pairing, also being created at the same time. The moral of this is to be careful what you train for, because our horses are always forming emotional (conditioned) responses to what we do with them. They're experiencing everything as either "good" or "bad" in one way or another and it's up to us to maximise the good experiences and minimise the "less good" ones, in order for learning to be effective. This is sometimes called "the behaviour bank account", and ours needs to be in a good credit position!

References

Baldwin, J. D., & Baldwin, J. I. (2000). *Behavior Principles in Everyday Life*. Pearson.

Busby, D., & Rutland, C. (2019). *The Horse a Natural History*. Quarto Publishing PLC.

Doherty, O., McGreevy, P. D., & Pearson, G. (2017). The importance of learning theory and equitation science to the veterinarian. *Applied Animal Behaviour Science*, *190*, 111–122.

Herrod, J. L., Snyder, S. K., Hart, J. B., Frantz, S. J., & Ayres, K. M. (2023). Applications of the Premack principle: A review of the literature. *Behavior Modification*, *47*(1), 219–246.

Mills, D. S. (1998). Applying learning theory to the management of the horse: The difference between getting it right and getting it wrong. *Equine Veterinary Journal*, *30*(S27), 44-48.

Mills, D. S., & Nankervis, K. (1999). *Equine Behaviour: Principles and Practice*. Blackwell Science.

Pearce, J. (2008). *Animal Learning and Cognition* (3rd ed.). Psychology Press.

Reid, P. I. (1996). *Explaining How Dogs Learn and How Best to Teach Them*. Excel-Erated Learning.

Chapter 5

Further learning theory

Further properties of classical conditioning

As we explored in chapter 4, classical conditioning can occur without human intervention. The horse simply learns that a particular neutral stimulus (e.g. car driving into the yard) is so closely associated with something (getting fed) that it triggers a conditioned response (kicking the door). In this section we explore what else can influence the learning process.

Salience

Up to now we have considered the relationship between stimulus and response as if they occur in isolation. But of course the stimuli that are used to elicit or cause a training response are in fact one of many competing stimuli in the environment. That's why the salience of the stimulus is so important. Salience describes the quality of being particularly noticeable, important or prominent. Consider that at any one time we are all bombarded by sensory input from sight, hearing, touch, taste and smell, as well as bodily conditions such as hunger, thirst and temperature (the stimulus collage that was introduced in chapter 4). Generally we ignore, or in behavioural terms become habituated to, most sensory inputs as they have little impact on our immediate wellbeing. Instead, we tend to focus our attention on a narrow band of currently important information whilst remaining alert to any wider changes that might highlight a potential threat or opportunity. Our horses are no different. One of the reasons clicker training can be so effective is that it uses a sound that the horse is unlikely to come across outside the training environment. This makes the noise more salient or noticeable against the background of other sensory inputs and therefore more likely to influence the horse's behaviour. Other short sharp noises, for example the expression "a-ah", are salient stimuli that you have probably seen used effectively to interrupt an undesirable response. The key is for the stimulus to be quick to administer, meaning it can be used with high contingency to mark the target behaviour and thereby allow detailed training (Baldwin and Baldwin, 2000).

The importance of salience in training cannot be understated. Horses have very different sensory capabilities to us and tend to perceive their world quite differently. We need to be sufficiently aware about the sensory inputs that are salient to the horse in order to ensure our cue stimulus is what the horse pays attention to, rather than other environmental stimuli.

DOI: 10.1201/9781003581390-7

Interestingly there is a psychological phenomenon in humans called inattentional blindness. Research has shown that if attention is being paid (sufficiently) to one thing, other things will not be picked up by our attention processes. One particular study, known as the invisible gorilla (Drew *et al.*, 2013), involves observers of a ball-passing game being asked to count the number of passes between players and report back their count at the end of the game. While the game is taking place a person in a gorilla suit wanders through the game. Despite having walked several times through the centre of the scene, the gorilla is not reported or even noticed by a large proportion of the observers. This is because when you are engaged in a demanding task requiring high levels of concentration (counting the number of passes), attention can act like a set of blinkers, making it possible for salient stimuli to pass unnoticed right in front of our eyes. The same phenomenon is likely to happen with horses. Consider the horse so fixated on something happening outside the arena they are unable to attend to the cues being given by the trainer. This is particularly true if the horse perceives whatever is going on as potentially threatening. We must be alert to the potential for inattentional blindness in our horses and take steps to ensure that whilst our horse is acquiring new learning, our cues are salient, for example by scheduling the training to a time and place where other potentially competing stimuli are minimised. This is part of the "set your horse up to succeed" mindset. The other important aspect of salience is that some stimuli or experiences carry greater or lesser importance or value. From the horse's perspective, grass is often more salient than some type of food rewards, which can be a challenge when training with food in the field!

Contingency

Contingency relates to the close and predictable relationship between the behaviour (e.g. a horse touching a target with their nose) (Figure 5.1), the secondary conditioned reinforcer (the clicker) and the primary reinforcer (food or scratch) (Mills, 1998). The contingency in this process is that the reward is contingent upon the desired behaviour (touch the target); the horse needs to touch the target in order to receive the reward. Put another way, the horse learns that receiving the reward is a consequence of touching the target. The clicker sound acts as a bridge, linking the behaviour to the reward. This clear, immediate linkage helps the horse understand which specific behaviour is being rewarded.

Figure 5.1 Aliyah Woodland of the Natural Animal Centre (NAC), target training her horse Bee. Photograph by Benny Payne (B. V. Productions).

Blocking

In blocking, the conditioning of a valuable predictive stimulus is hampered by the strength of a conditioned stimulus established by prior conditioning (Reid, 1996). What this means in practice is that a horse may have already learned a response to a cue which may be difficult to unlearn or to be replaced. Blocking occurs when a horse has been taught one cue (cue A), and then the trainer tries to associate the same behaviour (behaviour B) with a different cue. If the new cue is applied after the already learned cue, the horse is less likely to learn the new cue, because, in terms of predictability, there's no point, no need to make that association – they already know that cue A means do behaviour B. This can trip us up when we're trying to move from a verbal cue for walk on, to a leg cue for the same behaviour, for example. This is where the animal training maxim "new before old" comes from; it's better to give the new cue just before the existing learned cue, so that it provides the horse with information. They will learn that the leg cue (as yet unlearned) predicts the verbal cue (learned) and through classical conditioning the response to the leg cue will become conditioned. Then the verbal cue can be faded.

Overshadowing

Overshadowing is an important concept that helps explain how animals, including horses, learn to respond to one stimulus in the presence of several competing stimuli. Basic overshadowing occurs when the presence of a more salient or noticeable stimulus reduces the learning or conditioning strength of a less salient stimulus when both are presented simultaneously. For example, if a horse is being conditioned to respond to both the sound of a whistle and the tap of a whip at the same time, and the whip is more irritating than the whistle is loud, the horse might learn to respond more to the whip than to the whistle. This happens because the more noticeable stimulus (the whip) "overshadows" the less noticeable one (the whistle).

Robert Rescorla and Allan Wagner developed a more mathematical theory for understanding how overshadowing works alongside other phenomena like blocking (Miller *et al.*, 1995). According to the Rescorla–Wagner model, the amount of attention or learning capacity an animal can allocate to a stimulus is limited. Therefore, during the conditioning process, if two stimuli (e.g. the same whip and whistle as earlier) are presented together, the more salient stimulus (the irritating whip) will accrue more associative strength than the less salient whistle. According to this theory, the learning about the less salient stimulus is reduced not merely by its less intrinsic attention-grabbing properties but also by the amount of learning capacity already taken up by the more salient stimulus.

In contrast to the Rescorla–Wagner model, N. J. Mackintosh (1976) proposed an attentional theory of how learning can be overshadowed, by suggesting that the amount of attention paid to a stimulus is not fixed but can change over time based on the degree to which that stimulus predicts an outcome. According to Mackintosh, animals learn to pay more attention to stimuli that are better predictors of significant events (e.g. getting food or avoiding a whip tap). In the context of overshadowing, Mackintosh's theory suggests that overshadowing occurs because the more predictive stimulus comes to control more of the animal's attention, and thus, the less predictive stimulus is learned about less effectively. Consider this in the context of what is described as combined reinforcement, where a horse gets a whip tap for moving their hindquarters away from the mounting block, for example, and receives some food when they stand lined up at the block. Any guesses as to which way the direction of overshadowing might go there?

Understanding overshadowing can greatly assist in designing effective training programmes or behaviour treatment plans for horses, and it also helps us to understand why an owner's expected training outcomes sometimes fail to materialise. Practitioners need to consider which stimuli are most likely to capture the horse's attention, and how potentially competing stimuli might interact. For example, when training a horse to respond to a voice cue, you might need to ensure that environmental distractions don't overshadow the auditory cues. Recognising how horses allocate their attention to different stimuli can help us manage and modify equine behaviour more effectively.

Single event learning

Single event learning, often referred to as one-trial learning, is a form of associative learning where a significant event leads to the formation of a strong memory after just one experience (Figure 5.2). This type of learning is crucial in situations where the ability to learn quickly from one incident can be life-saving, such as avoiding predators or hazardous environments.

The famous Little Albert experiment, conducted by John B. Watson and Rosalie Rayner, exemplifies how emotions such as fear can be conditioned in humans, and this has been widely extrapolated to other species. In this experiment, a young child, Little Albert, was conditioned to fear a white rat (initially a neutral stimulus) by pairing the rat with a loud, scary noise (an unconditioned stimulus) that naturally elicited fear (an unconditioned response). After a few pairings, Little Albert began to show fear of the rat alone. Interestingly, although this experiment involved repeated exposures, it is claimed as evidence of single event learning in that a particularly strong or traumatic pairing can lead to a long-lasting behavioural change after just one or few trials.

Figure 5.2 Even if this was this horse's first fearful experience in an arena, this can set up a conditioned fear response resulting in behaviour such as refusing to enter the arena on the next occasion. Photograph by Uzo Borewicz.

In horses, single event learning can be particularly evident in cases of traumatic experiences. For example, if a horse is involved in a trailer accident, the traumatic experience might cause a conditioned emotional (fear) response, in other words a lasting aversion to the trailer. Behaviourally, they might refuse to enter a trailer again or display avoidant signs when approaching the trailer. This is because of the association created between the trailer and intense fear and possibly pain experienced during the accident, and this single event is enough to create a powerful, enduring memory.

Revisiting an evolutionary perspective, such learning is crucial for survival in the wild, as it enables horses to avoid threats and dangers after just one exposure. In domestic settings, it underlines the importance of ensuring positive or at least neutral first experiences when introducing horses to new situations, objects or tasks. Negative experiences can lead, sometimes after only one or two exposures, to long-term aversions that can be challenging to rehabilitate and can affect the horse's training and welfare.

Generalisation and discrimination

An adaptive feature of learning is that horses are able to both generalise and discriminate between stimuli or cues (Baldwin and Baldwin, 2000). For example, a riding school pony has to respond appropriately to many different riders using leg, seat and hand cues in different ways. The adaptive response of the pony is to generalise and respond in the same way to a broad range of varying but similar cues. In contrast an advanced level dressage horse has learned to discriminate between specific cues that indicate which response is required. This might mean small but unique adjustments to the rider's hands, leg or seat position to indicate sideways, forward or elevated movement.

In simple behavioural terms, stimulus generalisation occurs when the horse has learned a cue (conditioned stimulus) such as an increase in leg pressure means go forward, and they then generalise this response to any sufficiently similar stimuli, such as legs flapping or banging or wiggling. In effect any movement of the rider's leg will elicit the forward response. However, the stimulus that is the most similar to the conditioned stimulus will elicit the strongest response and those less similar will elicit the weakest response.

These examples illustrate how these learning principles operate differently depending on the level of training and the specific demands placed on the horse.

Generalisation allows horses to apply learned responses to similar stimuli, which is beneficial in many practical situations, as in the riding school pony example who must adapt to various riders who may not always apply cues consistently or precisely. Generalisation helps the pony maintain a level of predictability and safety in their responses, despite varied inputs. Discrimination, on the other hand, involves the ability to perceive and respond differently to distinct stimuli. This is crucial for advanced training, such as in dressage, where precise movements are required. The dressage horse learns to differentiate between very subtle differences in cues, which allows for a higher level of performance and more complex behaviours.

The intensity of the stimulus and the similarity both influence the strength of the response in the context of generalisation. This concept is a key point in understanding how horses respond across different contexts where we expect them to learn and adapt to our cues.

Horses can learn to discriminate between several different stimuli. Consider the horse who will confidently enter a trailer during practice sessions to get a bucket feed, but come show day, that same horse may refuse to load for that same reward. In this case the discriminative stimulus picked up by the horse may not be immediately obvious. Perhaps the addition of the tail bandage and boots or an unusually stressed owner on show day elicited this change in response. Stud farms can use the horse's ability to discriminate between cues to help a stallion differentiate between being brought out to cover a mare and being brought out to be trained, turned out or groomed. The consistent and repeated use of a leather headcollar when being led out to cover a mare results in the stallion getting sexually excited as soon as this halter is used. In contrast if a nylon halter is used repeatedly and consistently when the stallion is being led out for training, turnout or grooming, he learns to discriminate between the two different types of equipment and their association with what happens next.

Discrimination and generalisation, taken together, result in stimulus control (Reid, 1996). A horse can be trained to differentiate between cues very precisely and then learn to generalise their response so that it occurs in different locations, for different people and for different rewards.

Second-order conditioning

Second-order conditioning (also known as higher-order conditioning) refers to the process where a neutral stimulus elicits a response *not* by being directly paired with the unconditioned stimulus but by being paired with another conditioned stimulus (Baldwin and Baldwin, 2000). Just to recap:

- An unconditioned stimulus (US) is something that happens in the environment (tractor driving past field) that elicits an unconditioned response (UR) which is natural and automatic (running away).
- A conditioned stimulus (CS) is where a previously neutral stimulus (car driving onto the yard) after being repeatedly paired with the US (being fed) comes to elicit a conditioned response (door kicking).

Taking this example a step further, if the car driving onto the yard is repeatedly associated with food – perhaps the owner arrives morning and evening by car and then immediately feeds – their horse may eventually start banging the door (or field gate) just to the sound of cars arriving even without the presence of food. Second-order conditioning demonstrates the power of association in learning. A stimulus can acquire the ability to elicit a response based on its association with another stimulus, not just direct experiences. It also emphasises the intricate ways in which our and our horses' brains can form and understand associative chains. Using the same example, supposing every day just before the car arrives on the yard, next door's dog starts barking. If this new neutral stimulus (dog barking) is paired repeatedly with the car arriving (CS1) then, over time, the dog barking (CS2) will come to elicit the same conditioned response (door kicking) even though it was never directly paired with food (US).

Clicker training is often considered an example of second-order conditioning, although it initially begins as a straightforward case of classical conditioning.

Here's how it works:

First-order conditioning: in the initial phase, the clicker sound (neutral stimulus) is paired with a reward (unconditioned stimulus), such as food, which naturally elicits a response (e.g. a pleasant feeling). After repeated pairings, the clicker sound itself becomes a conditioned stimulus, capable of eliciting a conditioned response similar to the response to the reward.

Second-order conditioning: once the clicker has been established as a conditioned stimulus associated with the primary reinforcer of food, it can be used to condition further behaviours. For example, a trainer might use the clicker to mark a specific behaviour (like touching a target). In this phase, the clicker, already a conditioned stimulus, is used to build new associations. Here, the clicker (CS1) has been linked to the food reward (US), and then the behaviour (like touching the target, which becomes CS2) is also linked to the clicker. The clicker marks very clearly what behaviour the horse did, that will very soon result in a food reward.

This layering of associations – using a conditioned stimulus from one pairing to establish another conditioned stimulus – is what characterises second-order conditioning. In practical terms, the clicker's sound comes to not only predict food directly but also to reinforce and increase the likelihood of behaviours that have been marked by the clicker sound in training sessions. This makes clicker training a powerful method for shaping complex behaviours in training horses.

Extinction

Extinction refers to the gradual weakening and eventual disappearance of a conditioned response. If a behaviour that has been previously rewarded is no longer reinforced then over time the behaviour is likely to decrease in frequency or stop altogether (Reid, 1996). An extinction schedule is the opposite of continuous reinforcement and means that no responses are reinforced. A behaviour that is not reinforced in any way will tend to stop being offered. If the horse is expecting to be rewarded then the lack of reinforcement may initially lead to an increase in the performance of the behaviour. Generally over time, if the behaviour is not reinforced, it will decline in frequency although there may be episodes of spontaneous recovery where the horse offers the behaviour "just in case" they get the reward they're after. For instance, a horse that tends to mug their owner for treats may be allowed to mug repeatedly without the reward of receiving a treat. Initially the mugging behaviour may worsen but then will gradually reduce as the horse learns that mugging doesn't achieve their desired outcome. In other words they get nothing for their effort.

Example 1: Positive reinforcement extinction

Imagine you are training a horse to touch a target with their nose. Every time the horse touches the target, you give them a treat. The horse learns that touching the target results in a treat and therefore continues the behaviour. If you stop giving treats when the horse touches the target, over time, the horse will stop touching the target because the behaviour is no longer being reinforced.

Example 2: Negative reinforcement extinction

When lunging a horse some trainers might use a whip, not to hit the horse, but with the intention of encouraging movement by waving it or letting it trail on the ground behind the horse. The horse moves to avoid the sensation or the perception of the snake-like whip trailing near them, finding relief when they move in the desired direction. If the handler keeps waving the whip even after the horse has started moving correctly, without providing relief (i.e. ceasing the waving action), the horse stops responding to the cue of the whip because the expected removal of the whip stimulus (negative reinforcement) does not arrive.

Example 3: Punishment extinction

If a horse nips or bites and is immediately tapped on the muzzle every time as a form of punishment, they might learn that nipping leads to an unpleasant consequence and therefore the nipping behaviour decreases. If the horse starts to nip but no longer receives the tap on the muzzle, over time, the horse might go back to nipping as the behaviour is no longer being punished.

It's important to recognise that when the horse has learned to expect positive reinforcement for a certain behaviour but the positive reinforcement (provided by an appetitive stimulus) ceases, there may be an initial increase in the behaviour, known as "extinction burst", as the horse attempts to get the desired outcome. If a horse is used to getting a food reward for performing a specific behaviour and the handler stops rewarding the horse with a treat, they may attempt the behaviour more intensely or more frequently before it eventually ceases.

While extinction is a valuable tool in behavioural modification, it should be used thoughtfully and in combination with other training methods for the best outcomes. Also, the principle works best when the extinction process is consistent. If you sometimes reinforce the behaviour and sometimes don't, you are effectively putting it on a variable schedule of reinforcement, which can make the behaviour even more resistant to extinction.

Extinction in classical conditioning

In classical conditioning, an unconditioned stimulus (which naturally elicits a response) is paired with a neutral stimulus until the neutral stimulus alone can elicit the same response, thus becoming a conditioned response. Extinction, in this context, refers to the weakening and eventual disappearance of the conditioned response when the conditioned stimulus is repeatedly presented *without* the unconditioned stimulus (Baldwin and Baldwin, 2000). Over time, if the association between the two stimuli isn't maintained, the conditioned response will diminish.

Example 1: Bell and feed association

Suppose every time you fed a horse you rang a bell just before you put the bucket in the stable. Initially, the bell (neutral stimulus) means nothing to the horse. But over repeated pairings, they begin to associate the bell with the forthcoming feed (unconditioned stimulus). Eventually, the horse might start showing signs of excitement, such as kicking the stable door, merely at the sound of the bell, even before the feed appears. This new response to the bell is the conditioned response. If you started ringing the bell repeatedly without subsequently offering the feed, the horse will eventually stop associating the bell with the feed. Over time, the sound of the bell will cease to elicit the door-kicking behaviour, as the conditioned response undergoes extinction.

Example 2: Saddle and pressure

If a horse first wears a badly fitting saddle, they might feel slight pressure or discomfort (unconditioned stimulus), making them tense up (unconditioned response). If this happens every time they wear the saddle (neutral stimulus), they may eventually tense up (conditioned response) just at the sight of the saddle (which has now become a conditioned stimulus). If the horse wears the saddle repeatedly but doesn't experience the discomfort or pressure (due to adjustments made so that it fits better), the anxiety or tension they felt merely at the sight of the saddle may begin to wane, leading to extinction of the conditioned response.

Example 3: Farrier visits and anxiety

A horse has had a painful experience with a farrier when they were younger. The sight of the farrier's tools (neutral stimulus) has become associated with the pain or discomfort (unconditioned stimulus) and thus elicits attempts to escape the situation (conditioned response) in the horse whenever they see the tools. If the horse has several positive and pain-free experiences with the farrier, over time, the escape attempts at the sight of the tools will decrease. By consistently seeing the tools without any negative outcome, the conditioned response can undergo extinction. However, the process to achieve this may need to be carefully thought out, more explanation of this in the chapter on behaviour modification.

Spontaneous recovery

It's crucial to understand that even after extinction, there might be instances of "spontaneous recovery" (Baldwin and Baldwin, 2000). This phenomenon refers to the sudden reappearance of the conditioned response after a period of time, even without the presence of the unconditioned stimulus. For instance, even if the horse has stopped kicking the stable door at the sound of the bell, after some time, the sound might again elicit that behaviour. Over successive instances, these spontaneous recoveries will reduce in intensity if the extinction process remains consistent.

Flooding

In order to understand flooding it's worth recapping aspects of learning we've looked at previously in chapter 4 – habituation and sensitisation. We've established that a horse's response to a stimulus gradually diminishes through many repetitions as the horse "gets used to things". However, if the stimulus (the thing) is fear provoking and the horse is unable to get far enough away to feel safe, then a psychological effect known as "flooding" is more likely to occur (Reid, 1996). Flooding is a controversial process as it involves the horse being forced to confront a fearful situation until they no longer respond in a fearful manner. What happens when we overwhelm our horses in this way is that we can make them more scared of things rather than less scared. In this scenario it's easy to see how horses can develop behaviour that is labelled as flighty or spooky. They have become sensitised to the things we wanted them to be chilled about. A familiar human example of flooding is where someone with a fear of spiders is put into a room full of spiders with no means of escape until they appear to be no longer afraid (if only it was that easy!). With horses, the situation may typically be where a horse is forced, usually by restraint on a lead rope or because they are in a stable and can't escape, to be much closer to something scary than they want to be (Figure 5.3). This can go one of two ways; flooding can be effective and the horse then "habituates" to the thing they found scary, or the horse can develop a more extreme fear or phobia of the scary thing. Flooding is a high risk strategy because it's not possible to say which way it will go for an individual horse and even when a horse appears to habituate to, or appears to be ok about, the scary thing, there may be an unseen cost of increased anxiety or stress that will show up elsewhere in their behaviour.

Flooding is not an ethical practice since it is amplifying the horse's fear response due to the lack of escape, their lack of control and the repeated intense exposures. For example, a horse who is fearful of the hose is tied up (no opportunity to escape) and hosed down (flooding) until they eventually stop trying to get away. They appear to be "cured" and less fearful of the hose in future as they no longer try to escape being hosed down. Note the word "appear"; the absence of a fear response does not mean that the fear has disappeared. The internal physiological signs of fear are still present.

In human clinical psychology, the informed consent of the patient has to be obtained in order for this technique to be used. We can't ask a horse to give their informed consent!

Figure 5.3 Flooding: inescapable fear of a painful rope halter. Photograph by shymar27.

Learned helplessness

Continuing with the ethics of enforced fear, when a highly aversive stimulus is inescapable then learned helplessness may occur (Hall *et al.*, 2008). In effect, the horse no longer tries to escape; they simply give up and become dull to the experience (as in the previous hosing down example). The horse has learned that their behaviour makes no difference to the aversive environment and that they have no control over what is happening. In behavioural terms we can say that noncontingent consequences that occur at high frequency can condition passivity. If you remember from earlier in the chapter, high contingency relates to the close relationship in time between the behaviour and the reinforcer so that they are linked. Noncontingent consequences therefore relate to the lack of a relationship between the consequence and the behaviour. This is particularly pertinent in noncontingent punishment as the horse cannot be clear about what they need to do (or not do) to stop the punishment. People who receive frequent noncontingent punishment can develop learned helplessness associated with passivity, apathy and depression (Maier and Seligman, 1976). When people are first exposed to this type of consequence they try to avoid it or stop it; however, after repeated attempts to avoid or stop the punishment have failed, the person stops "fighting" and learns to accept the pain/discomfort passively. It is likely that this too occurs in horses. Once a horse has learned that their behaviour (whatever they do) does not provide any escape from an aversive situation, then in future when presented with other aversive situations they simply make no attempt to escape. Such horses appear to be shut down, and their ability to learn new things is restricted as they are unwilling to experiment (trial and error) with their behaviour to find the "right" response.

References

Baldwin, J. D., & Baldwin, J. I. (2000). *Behavior Principles in Everyday Life*. Pearson.

Drew, T., Võ, M. L. H., & Wolfe, J. M. (2013). The invisible gorilla strikes again: Sustained inattentional blindness in expert observers. *Psychological Science*, *24*(9), 1848–1853.

Hall, C., Goodwin, D., Heleski, C., Randle, H., & Waran, N. (2008). Is there evidence of learned helplessness in horses? *Journal of Applied Animal Welfare Science*, *11*(3), 249–266.

Mackintosh, N. J. (1976). Overshadowing and stimulus intensity. *Animal Learning and Behaviour*, *4*.

Maier, S. F., & Seligman, M. E. (1976). Learned helplessness: Theory and evidence. *Journal of Experimental Psychology: General*, *105*(1).

Miller, R. R., Barnet, R. C., & Grahame, N. J. (1995). Assessment of the Rescorla-Wagner model. *Psychological Bulletin*, *117*(3), 363.

Mills, D. S. (1998). Applying learning theory to the management of the horse: the difference between getting it right and getting it wrong. *Equine Veterinary Journal*, *30*(S27), 44-48.

Reid, P. I. (1996). *Explaining How Dogs Learn and How Best to Teach Them: Excel-Erated Learning*. James & Kenneth Publishers.

Chapter 6

Physiology of behaviour

The importance of understanding equine neuropsychology

Equine neuropsychology is the study of the nervous system's role in equine behaviour and cognition. Understanding the biological basis of behaviour provides critical insights into what is happening in the brain and how that impacts on the way the horse is feeling and behaving. It allows us to generate better explanations for why behaviour is taking place and come up with more effective solutions for training and behaviour modification.

What we need to know about the horse's brain

The horse's brain weighs around 700 grams in comparison to the human brain which is roughly double the size at 1,400 grams. From a structural point of view they are fairly similar, the main difference being in the prefrontal cortex which is located in the frontal lobe of the cerebral cortex. The prefrontal cortex of the horse is significantly smaller and takes up a smaller percentage of the brain as a whole compared to the human equivalent. Why is this important? Well, we refer to the prefrontal cortex as the executive centre where important decision-making and planning takes place. It's also responsible for impulse control, regulating emotion, thinking and solving puzzles. So now we can see the significance of the difference in size between the horse and the human in this area of the brain. Humans, with their relatively large prefrontal cortex, have the ability to suppress reactive responses, whereas horses are more likely to act on them. This is critically important to bear in mind when considering how we manage, train and work with our horses. Our horse may not have much choice in responding to the scary, flapping thing in the hedge by turning and galloping in the opposite direction.

An understanding of the prefrontal cortex is helpful, as is, as far as behaviour control is concerned, the set of structures collectively referred to as the striatum (Hemmings, 2023). The striatum is made up of the cordatus, the putamen and the nucleus accumbens. The cordatus and the putamen are particularly instrumental in fine motor control and they also play a crucial role in the formation of habits – those behaviours that when practised over time become automatic.

The nucleus accumbens is particularly crucial in the early stages of learning a new task, particularly if we use positive reinforcement in the form of a food reward. This area is activated when the horse (or we) experience something pleasurable. So courtship and mating are also

DOI: 10.1201/9781003581390-8

key activators. This activation is triggered by the release of the neurotransmitter dopamine. Dopamine is integral in a wide range of behaviours, being released in the brain during rewarding experiences. This provides the motivation to seek out those rewarding behaviours and making us (and our horses) more likely to repeat them. In this way dopamine plays a central role in reinforcement learning by connecting behaviours with pleasure and reinforcing the link between cues and upcoming rewards. However, it also has a role as a stress neurotransmitter, and research has shown a significant positive correlation between anxiety and spontaneous blink rate, a measure of dopamine release. Worth remembering when we consider stereotypies later.

Neural pathways

A neural pathway is a series of connected neurons (or nerve cells) which send and receive signals from the brain. The connection between neurons are called synapses. A synapse is the tiny gap across which a neuron can send an impulse to another neuron (Figure 6.1). Brain activity involves both electrical signalling (action potentials) and chemical signalling (neurotransmitters) (Toates, 2007). Neurons talk to each other and send chemical messages back and forth in response to electrical impulses; chemical messages take the form of neurotransmitters such as dopamine. This cycle of electrical and chemical activity is fundamental to all functions of the nervous system. The connections between neurons form specific pathways in the brain: if a skill or behaviour is practised repeatedly the connections between the neurons become strengthened and form a neural pathway. Imagine these neural pathways like water flowing down a mountain; they find a route and over time the water carves a path for itself in the rock, in other words the neural pathway is formed and consolidated. This helps to explain why behaviour modification can be difficult when a strong neural pathway has been formed. It takes a lot to divert the water from its well-worn (learned) route.

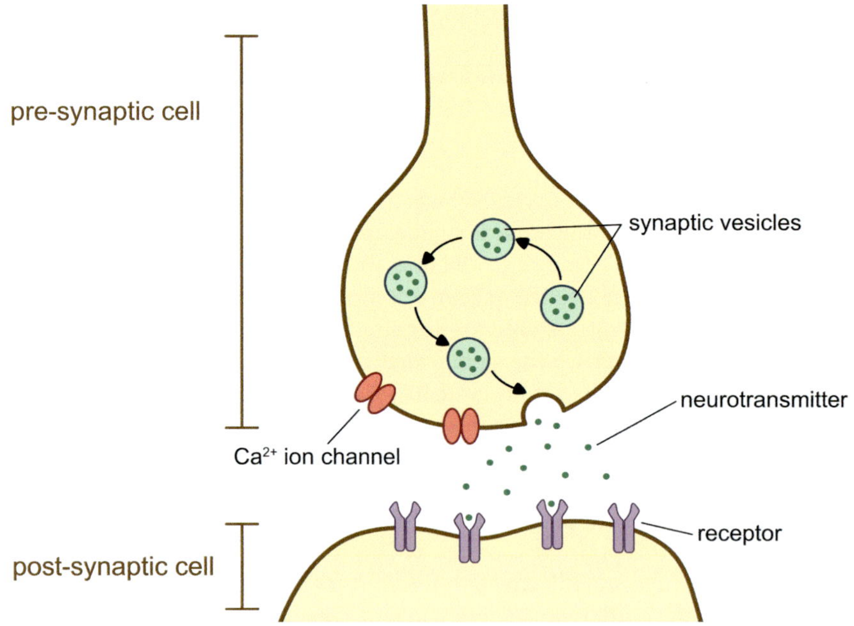

Figure 6.1 In synaptic transmission electrical and chemical signals reach the end of one neuron (the presynaptic neuron) and cross the gap (the synapse) to bind to receptors on the next (postsynaptic) neuron, thereby transmitting the signal onward. Diagram by aiyoshi597.

The development of neural pathways requires forming associations. To establish these associations, the cerebral cortex works in tandem with the striatum, supported by extensive reciprocal links between these areas that are crucial for developing learned associations. Dopamine plays a key role in reinforcing these new connections. For instance, in clicker training – a technique used to condition animal behaviour – the process starts when a horse hears a click sound. This sound is processed by the temporal lobe, which attends to auditory stimuli. Immediately after the click, the horse receives a food reward. This rewarding experience triggers the release of dopamine, which activates the nucleus accumbens within the striatum. If the process is repeated multiple times, the pathway of neurons that fire immediately before the reward is received becomes strengthened. That pathway relates to the temporal lobe that received the information, the click, in the first place. After a number of repetitions, the association between the audible click and the pleasurable experience (the food reward) is strengthened to the point that just hearing the click alone will fire up that neural pathway, causing dopamine to be released and "rewarding" the horse with a pleasurable feeling (classical conditioning, the click sound takes on the qualities of the food).

The Hebb rule

This pathway strengthening is referred to as long-term potentiation (LTP), "neurons that fire together, wire together" (Hebb, 1949). When neurons fire together repeatedly, their simultaneous activation strengthens the synaptic connections between them, enhancing the neural pathway. Conversely, long-term depression (LTD) involves the weakening of synaptic connections when activations are less frequent or asynchronous (not occurring at the same time). If neural connections are seldom used, this lack of activation leads to their inefficiency in transmitting signals. Over time, these underused connections are pruned – a process known as synaptic pruning. This helps in prioritising more frequently used pathways, thereby strengthening them. Synaptic pruning supports the concept of behavioural extinction by illustrating how less-used pathways can weaken over time, reflecting changes at the neural level that mirror behavioural adaptations.

Long-term potentiation can also be associative. Associative LTP occurs when two inputs (or stimuli), a weak one and a strong one, converge on a single neuron. If the weak input fires simultaneously with the strong input, the weak synapses can also undergo LTP and become strengthened. The benefit of this in training is that it allows for the learning of indirect associations by linking a new input to an already strong one.

The concept of "new before old" training aligns closely with the principle of associative LTP. In this training approach, you're effectively using the principles of associative learning to transfer a response from a well-established cue (an existing strong synaptic connection) to a new cue (a weaker synaptic connection).

This is how "new before old" would work in a situation where we are transferring a new cue to an existing cue that the horse knows for that behaviour. For example, a horse has learned to move off from a verbal cue "walk on" and the trainer now wants them to move off when the rider applies their leg gently to the horse's side:

- The horse has already learned to respond to the verbal cue "walk on", which is a strong cue that reliably elicits the behaviour of moving forward.
- By introducing the new cue (applying the rider's leg gently to the horse's side), simultaneously with or just before the known verbal cue, the horse begins to associate the new cue with the expected behaviour.
- Over time, with consistent and repeated pairing, the response to the new cue is strengthened through associative LTP. The neural pathways responsible for responding to the new cue become more potent, allowing the new cue to eventually elicit the behaviour on its own.

This method is effective because it leverages the existing strong connection (the verbal cue) to establish a new connection, making it easier and quicker for the horse to learn and adapt to the new cue. This approach is broadly used in various forms of horse training (whether the trainer realises it or not!) and in this example shows that responding to a leg aid doesn't always have to have been taught by applying the pressure in an aversive way via negative reinforcement.

In essence, when we're training an association, whether by operant or classical conditioning, we are actually developing a strengthened neural connection. Physical training and skill acquisition activate specific neural pathways repeatedly to induce LTP in those circuits, engraving the patterns like the flowing water we discussed earlier. This process involves neurobiological changes that alter the physical structure of the neural circuitry.

The limbic system: the centre of emotion and behaviour

The limbic system, sometimes referred to as the emotional brain, is a group of brain structures located underneath the cerebral cortex and is responsible for feelings, motivation and long-term memory, among other things. Structures within the limbic system play a key role in the processing and interpretation of emotional responses. The amygdala, for example, is particularly involved in processing fear, stress responses and pleasure. The limbic system is also involved in motivation, especially areas like the hypothalamus and the nucleus accumbens. These regions are integral to the reward circuitry of the brain, which is associated with motivational processes and pleasure-seeking behaviours. In terms of memory, the limbic system, particularly the hippocampus, is vital for the formation and retrieval of long-term memories, especially those related to individual experiences and spatial orientation. It's also interconnected with other brain regions that influence its functions; for instance, it works closely with the prefrontal cortex which impacts decision-making and moderates social behaviour.

Horses, like humans, have a limbic system that handles emotional processing. The amygdala's role in modulating fear is crucial for a prey animal, and the involvement of the hippocampus in forming and retrieving memories is vital for their learning and survival. Whilst they have the same structures in common, there are differences in complexity between the human and equine limbic systems: in humans, the interaction with the highly developed prefrontal cortex allows complex decision-making and problem-solving and moderates cognitive social behaviour. In horses, this area is less complex and the interaction doesn't play such a significant a role in these higher cognitive functions, although that's not to say that horses are not highly social animals, of course. Horses' behaviour and social interactions are driven more by the limbic system and less by the complex deliberative processes that humans engage in with their more developed prefrontal cortices (LeDoux, 1998).

Let's now look at some of the important structures of the limbic system in more detail.

Amygdala

The amygdala is critical for processing emotions, assigns emotional significance to sensory inputs and stores memories of fearful events which can shape future responses to similar stimuli. It is also the brain's alarm system, detecting threats and triggering fear responses such as heightened arousal, release of stress hormones (through interactions with the endocrine system), escape behaviour and freeze behaviours. When a horse encounters a potential threat, the amygdala processes this information and can trigger an immediate emotional response. This occurs before the cerebral cortex, which is involved in higher-order processing and impulse control, has a chance to fully assess and modulate the situation. The amygdala reacts both to environmental dangers and social threats. In stressful situations, the amygdala activates the hypothalamic-pituitary-adrenal (HPA) axis by sending signals to the hypothalamus. This triggers a cascade of hormonal responses, resulting in the release of adrenaline (epinephrine) from the adrenal medulla and cortisol from the adrenal cortex. These two stress hormones prime the body for rapid action. If they are not utilised in physical activities like running or fighting, they simply linger in the bloodstream. Consequently, to complete the stress cycle, it is necessary for the anxious horse (or human) to engage in movement. This activity helps use up the adrenaline and cortisol, thereby calming the amygdala's alarm system. The body is not designed to endure high levels of stress hormones for extended periods. Prolonged exposure can impair the growth of brain cells in the prefrontal cortex, adversely affecting learning and impulse control. For optimal functioning, the cortex requires a calm emotional state, while the amygdala benefits from being deactivated through emotional stability in relationships (secure attachment) and adequate sleep for the species.

The hormone oxytocin is released when we (or our horses) feel safe and connected in relationships. Oxytocin plays a crucial role in social bonding, maternal behaviours and trust-building among individuals. Horses get this social connection from their dam and from their herd or affiliative group, where synchronised movement helps them to feel safe, and also from ourselves as significant individuals in our relationship with them. In horses, oxytocin release can be stimulated by positive human interaction, gentle grooming, or any calm, bonding activity. The role of oxytocin is important to note in terms of how horses are typically managed, in terms of the stress that they experience when, under human control, they have to change turnout groups, their next door stable companion changes, or they are moved to another yard. Anything that disrupts their attachments with other horses or a relationally significant person negatively affects their feeling of safety.

Hippocampus

Also within the limbic system, the hippocampus is responsible for long-term memory formation and retrieval, as well as influencing mood and the emotional association in memories. This is important given that emotional memories (those linked with extreme pleasure, fear or trauma) are likely to be stronger and longer lasting than non-emotional memories. We'll consider memory and its role in behaviour later in the chapter.

Emotions play a crucial role in memory formation and retention, supported by these three functions:

1. *Enhanced encoding*: emotional arousal activates the amygdala, which we know is involved in emotional processing. The amygdala interacts with the hippocampus, crucial for forming new memories. This interaction helps to enhance the encoding of emotional memories, making them more vivid and detailed.
2. *Consolidation*: emotional experiences often lead to the release of stress hormones like adrenaline and cortisol, which can affect the consolidation phase of memory (the process by which short-term memories are converted into long-term memories). This hormonal boost can help to "seal" the memory, making it more durable.
3. *Retrieval*: emotional memories are often easier to recall than non-emotional ones. This is because the strong emotional context can serve as a powerful retrieval cue, bringing the memory to mind more readily.

Memories tied to strong emotions are not only more vivid but also more likely to be accessed and recalled over time compared to memories that lack an emotional component. In horses, a non-emotional memory could involve something routine and neutral, such as learning a specific trail route or the location of a valuable resource area. For example, a horse can learn the locations of various water sources around their pasture. This type of memory involves spatial and environmental cues, rather than emotional responses: the horse remembers these locations as part of their regular activities, without any significant emotional experiences tied to that memory.

Hypothalamus

The hypothalamus serves as a critical link between the nervous system and the endocrine system via the pituitary gland, and it regulates motivated behaviours such as eating, drinking and sexual activity through hormonal control. It also influences dopamine-mediated reward processing by interacting with areas such as the ventral tegmental area and the nucleus accumbens, which are integral to the limbic system. The hypothalamus also plays a key role in social bonding, recognition, and trust formation through its regulation of oxytocin release from the pituitary gland, and it activates the sympathetic nervous system in response to stress signals received from the amygdala.

Practical implications in training and handling

Understanding these emotional aspects of brain function can inform how we train and handle horses. Since their behaviour is less influenced by complex reasoning and more by instinct and past experiences (both good and bad), training methods that involve associative learning and habituation (or desensitisation if sensitisation has already taken place) are effective. The emotional state of a horse, such as fear or excitement, is more directly tied to immediate stimuli and past experiences than to intricate thought processes.

Introduction to the nervous system

The basic components of the nervous system are the central nervous system and the peripheral nervous system. The central nervous system (CNS) refers to those parts that are encased in bone. This means the brain covered by the skull and the spinal cord housed by the vertebral column. The brain and the spinal cord are connected to each other at the brain stem (the midbrain, the pons and the medulla oblongata). The brain stem is the oldest, most primitive part of the brain in evolutionary terms and its function is to keep you (and your horse) alive by coordinating functions that are essential for life and linking various parts of the nervous system.

The CNS integrates and coordinates the activities of the entire nervous system, while the peripheral nervous system (PNS) connects the CNS to the rest of the body. The PNS comprises all the nerves and neurons that reside outside the brain and spinal cord and is divided into two parts: the somatic nervous system (SoNS) and the autonomic nervous system (ANS). The SoNS controls voluntary movements and relays sensory information from the skin, muscles and joints to the CNS, encompassing both motor and sensory nerves. In contrast, the ANS regulates involuntary functions such as heartbeat, digestion and breathing. The CNS synthesises information and facilitates adaptive decision-making, while the PNS conveys signals between the CNS and the rest of the body to execute its commands. Together, they form the comprehensive nervous system that regulates all aspects of a horse's physiology.

The autonomic nervous system

The autonomic nervous system reacts to emotional states and stressors by activating either the sympathetic nervous system (responsible for fight, flight, freeze, fidget, and fawn responses) or the parasympathetic nervous system (which promotes relax, rest, and digest). The parasympathetic nervous system (PSNS) promotes relaxed physiological states conducive to rest, relaxation, and digestion. It slows the heart and respiration rates, increases blood supply to the gut and generally induces a calming effect on the body. In contrast, the sympathetic nervous system (SNS) has the opposite effect, accelerating bodily functions and preparing the body for action. It causes the heart rate to increase and the blood vessels supplying the gut to constrict, redirecting blood flow to the skeletal muscles in preparation for rapid movement. This system becomes activated in response to stress. Its primary function is to assist the horse in coping with acute situations rather than chronic ones, although it is activated in both. These physiological effects, also including prolonged high levels of adrenaline and cortisol as we saw earlier, help us to understand why chronic stress can have detrimental effects on a horse's wellbeing.

These ANS responses are influenced by factors such as attention, arousal, motivation and readiness to respond, and they can both enhance and impair cognitive functions, such as memory formation and recall. Research has indicated that while excess stress, or distress, can impair memory, moderate and manageable stress may actually enhance it. The state of the ANS also impacts learning capabilities: stressful situations often result in a narrow focus on threat-related cues, whereas relaxed states promote a broader focus, fostering openness to exploration and new learning.

The role of the vagus nerve and vagal tone

The vagus nerve is the cranial nerve that connects the brain to the body and is the longest and most complex of the cranial nerves. (Cranial nerves emerge directly from the brain, as opposed to spinal nerves that emerge from the spinal cord.) It carries sensory information about the state of the body's organs to the brain, as well as motor signals from the brain to the internal organs.

The autonomic nervous system is divided into two main branches: the sympathetic nervous system and the parasympathetic nervous system. The vagus nerve, also known as the tenth cranial nerve, serves as a major component of the PSNS, playing a crucial role in regulating many parasympathetic functions including heart rate, digestion, respiratory rate and various involuntary bodily functions. It acts as a primary pathway through which the brain communicates with organs and tissues to maintain homeostasis. It is also referred to as the "inflammatory nerve" because it serves as a vital communication pathway between the brain and the body, detecting inflammation and coordinating the body's response to it through the release of neurotransmitters such as acetylcholine.

Vagal tone refers to the activity of the vagus nerve at rest. Higher vagal tone indicates greater parasympathetic activation, resulting in lower heart rate, increased heart rate variability and quicker recovery from stress. These factors are linked to overall health, stress resilience and the ability to recover faster from stressors. Research has shown that high heart rate variability is associated with more efficient autonomic control, allowing greater responsivity and sensitivity to changing environmental demands (Tiwari *et al.*, 2021). This is because higher variability and lower resting heart rates reflect increased vagal action, which in turn is indicative of healthy autonomic functioning.

Training and conditioning, whether physical or mental, can help increase vagal tone and parasympathetic activation in horses, thereby improving their stress resilience. Low vagal tone is associated with chronic stress, high inflammation, increased disease risk and poor impulse control. Environmental stressors, transport stress, illness and injury can lower vagal tone, making the regulation of bodily processes less efficient. In essence, the vagus nerve and its activity level or tone play a crucial role in controlling many bodily functions and have widespread effects on overall health.

Neuroplasticity

Neuroplasticity is the brain's capability to undergo structural and functional changes in response to experiences and learning. This plasticity is crucial for processes such as learning, memory, recovery from injury, and the acquisition of new skills. Regular stimulation and learning promote neuroplasticity throughout life in both horses and humans. At the core of neuroplasticity lies synaptic plasticity, which refers to the ability of the connection between two neurons, called synapses, to adapt and change its strength over time.

A synapse is a specialised junction between two neurons, where electrical or chemical signals are transmitted from one neuron to another. Synapses can become stronger or weaker depending on their activity levels, playing a vital role in learning and memory processes. Long-term potentiation strengthens synapses when two connected neurons are repeatedly activated together, while long-term depression weakens unused synapses over time if activity is not sustained. The balance between LTP and LTD enables frequently used connections to be reinforced while unused ones diminish.

Synaptic transmission between neurons occurs when an electrical signal reaches the end of one neuron (the presynaptic neuron), triggering the release of chemical messengers called neurotransmitters. These neurotransmitters cross the small gap called the synapse to bind to receptors on the next neuron (the postsynaptic neuron), thereby transmitting the signal onward.

A study by Henshall *et al.* (2022) found that neuroplasticity and the acquisition of learning in horses may be improved by exercise prior to training. A brief warmup where the horse moves calmly through all paces reduced the time to learn a new task during subsequent training. In contrast, exposure to uncontrollable stress caused high cortisol release and hindered learning. The study found that horses with high cortisol levels required more trials to learn a new task. What is perhaps surprising about the high cortisol group was that they were the inactive control group who were tied up for 35 minutes with a companion and no other interventions prior to attempting the learning task. The researchers speculated that the high cortisol levels were due to the restraint which limited the horse's behavioural choices and increased their level of stress. These findings, and those of other studies that found that learning is impeded under high stress, are very important to bear in mind when we are planning training or behaviour therapy programmes. Consider the salience of restricted movement to the horses in the aforementioned study and why such restraint might have been so stressful for them (survival depends on being able to run away), and it's easy to see how different a human's perspective of a stressful environment is, compared with a horse's. This is just one demonstration of why it's more helpful to base our management and training decisions on what we know about a horse's ethology and their physical and emotional needs, rather than assumptions taken from our own lives as a different species.

The impact of stress

At this point it's worth defining what we mean by stress in the context of the physiology of behaviour. Stress is an experience, context or stimulus perceived to threaten or disturb homeostasis (the optimal physiological basis of wellbeing). It encompasses a state of heightened arousal, a perception of aversiveness and a lack of control. At the cognitive level, stress may induce a state of uncertainty about the most appropriate action to take to safeguard physical, emotional or mental wellbeing. At a manageable level, stress exposure can be adaptive, causing beneficial modifications to brain regions to support learning; however, it's maladaptive when it can't be escaped or controlled.

There are two main routes by which the body responds to stress – the sympathetic nervous system pathway and the hypothalamic-pituitary-adrenal (HPA) axis pathway (Figure 6.2). These two pathways work together when faced with stressors. The SNS provides an instantaneous response, while the HPA axis sustains the energetic and metabolic needs to deal with extended challenges. Both are evolutionarily adaptive, meaning they have developed over time in order to improve the chances of survival, but chronic activation can be detrimental to health.

When a potential threat is detected, the amygdala sends out a distress signal and within seconds the sympathetic nervous system is activated, triggering the release of adrenaline and noradrenaline. This increases heart rate, blood pressure, breathing rate, metabolism and muscle tension, preparing the body for immediate action. In tandem, although over minutes rather than seconds, the hypothalamus triggers the production of cortisol which increases available energy by mobilising glucose and fatty acids whilst suppressing inflammation and the immune response. The cortisol levels in the blood start to rise within five to ten minutes while the higher blood sugar and suppressed immune response it causes builds up over hours. The HPA axis operates more slowly than the SNS in order to deal with the longer term physical impacts of stress. If it remains activated for too long as a result of chronic stress it can lead to:

- decreased immune function and higher susceptibility to illness
- elevated cortisol which over long periods of time can contribute to laminitis, obesity and insulin resistance. Worth bearing in mind this link to metabolic compromise; laminitis isn't all about the grass!
- disruption of the reproductive hormones testosterone and progesterone, impacting negatively on fertility
- chronically elevated noradrenaline, adrenaline and dopamine levels which contributes to anxious or aggressive behaviour and hyperreactivity
- impaired decision-making, learning and training due to the stress effects on the brain
- the parasympathetic nervous system being unable to properly relax the body following a stress event

This all highlights the importance of providing a low-stress handling and living environment in order to prevent the chronic activation of the stress response and the subsequent impact it has on the body systems and health of the horse.

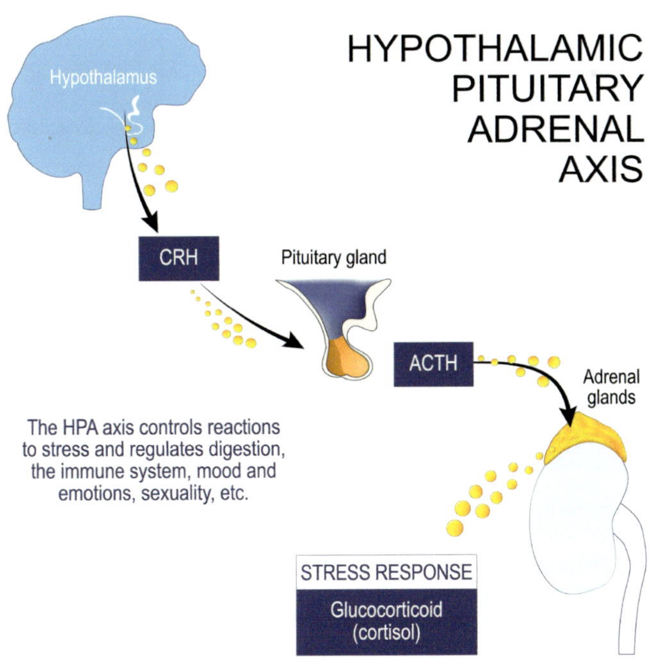

Figure 6.2 The HPA (hypothalamic-pituitary-adrenal) axis. Diagram by Designua.

Sleep: the cornerstone of physiological function

Sleep is divided into several stages, typically categorised into non-rapid eye movement (non-REM) and rapid eye movement (REM) sleep (Table 6.1). Non-REM sleep is further divided into three stages:

- *Stage N1*: This initial stage of sleep is light and easy to wake from. It serves as a transition from wakefulness to sleep.
- *Stage N2*: Sleep deepens in this stage; it's harder to awaken, and bodily functions like heart rate and breathing slow down.
- *Stage N3*: Known as deep or slow-wave sleep (SWS), this stage is crucial for physical restoration and recovery. It is the most rejuvenating and the deepest sleep stage, involving significant physiological changes in the body.

REM sleep, which follows non-REM stages in a typical sleep cycle, involves active brain function but also muscle paralysis (except for muscles that control eye movements and breathing). This stage is associated with vivid dreaming in humans and is thought to support brain functions like memory consolidation and emotional regulation.

Table 6.1 Sleep Stages Observed in the Horse

Sleep stages	Characteristics
Transition between wakefulness and sleep	The horse is drowsy. Lowers head intermittently. Eyes open and close slowly. This stage is the lightest form of sleep, marking the initial transition from wakefulness to sleep.
Light sleep	Muscle tone is still present, allowing the horse to stand, with the aid of the stay apparatus. Breathing remains regular. This stage provides rest but horses remain easily awakenable, allowing for quick responses to environmental changes.
Deeper sleep	The horse begins to sleep more soundly, potentially including snoring. Reduced physical movement. Heart rate and body temperature decrease. Brief powerful bursts of neuronal firing (sleep spindles) believed to be linked to synaptic plasticity. Sleep spindles play an important role in memory consolidation.
Deepest non-REM sleep	This is the most restorative sleep stage for horses, crucial for tissue repair, muscle and bone growth, and immune system strengthening. Procedural and motor memories are consolidated during this phase. Horses typically need to lie down in this stage to achieve the full benefits of deep sleep.
REM (paradoxical) sleep	Loss of muscle tone is accompanied by rapid movement of the eyes as they dart back and forth under closed eyelids. Lateral recumbency (lying flat out) – it is not possible for horses to experience REM sleep standing up. Sternal recumbency with muzzle touching the ground. Heart rate and breathing can be irregular. Brain activity levels are high, resembling those during wakefulness. REM sleep is crucial for processing emotional memories and complex brain functions, despite being less physically restorative.

Transitions between sleep and wake states are orchestrated by multiple brain structures, each with a unique role:

- *Hypothalamus*: central in controlling the onset of sleep and wakefulness. It regulates various bodily functions that are crucial for maintaining sleep cycles, such as temperature and hormonal balance, often through interactions with the pineal gland which produces melatonin.
- *Hippocampus*: active during sleep, particularly REM sleep. It processes and consolidates memories, integrating new experiences with existing memory networks.
- *Amygdala*: as the emotional processing centre of the brain, the amygdala is particularly active during REM sleep, supporting the processing of emotional experiences and memory integration.
- *Thalamus*: acting as a gatekeeper for sensory information, the thalamus blocks most sensory signals to the cortex during non-REM sleep, helping to maintain the sleep state, and becomes active during REM sleep.
- *Reticular formation*: located in the brainstem, this network of neurons is crucial for regulating the sleep-wake cycle. It helps maintain cortical alertness and is essential in the transition between sleep and wake states.
- *Pons*: This part of the brainstem plays a key role in initiating REM sleep. It sends signals to relax muscles and inhibit movement, ensuring the characteristic muscle atonia of REM sleep.

From the previous we can see that sleep is a behaviour, it's not just a time when a horse is switched off and doing nothing. The primary value of sleep is believed to be the restoration of balance among neural centres which are necessary for overall health, and it's certain that sleep deprivation compromises health significantly. However, the specific physiological functions of sleep are still not fully understood and are the subject of ongoing research. Current hypotheses suggest several critical roles of sleep, including the following:

1. *Facilitation of learning and memory*: sleep is crucial for the consolidation of memories, helping to stabilise and integrate new information learned during wakefulness into long-term memory.
2. *Synaptic homeostasis*: sleep may involve the targeted weakening or pruning of synapses – this "synaptic downscaling" is thought to help "forget" less important information, thus preventing the overload of neural circuits and maintaining synaptic efficiency.
3. *Clearance of metabolic waste*: during sleep, the brain clears out metabolic waste products that accumulate during periods of wakefulness. This process is facilitated by the glymphatic system, which increases in activity during sleep to remove toxins that could potentially impair brain function.

(Greening and McBride, 2022; Patel *et al.*, 2022)

Sleep cycles are regulated by the animal's circadian rhythm (Aleman *et al.*, 2008). Circadian rhythms are physical, mental and behavioural changes that follow a 24-hour cycle. These natural processes respond primarily to light and darkness. We know that horses are crepuscular, meaning they are most active at dawn and dusk. We also know that they have polyphasic sleep patterns, meaning they have intermittent periods of rest and sleep throughout a 24-hour period. They also require far less sleep than humans, averaging a total of only three to five hours per day. Most of their sleep happens at night, particularly if they are stabled. Deep sleep in horses is short in nature with five minutes of slow-wave sleep, followed by five minutes of REM sleep and then another five minutes of slow-wave sleep, and can only be achieved in lateral recumbency, or in sternal recumbency with the muzzle touching the ground.

Memory consolidation

During sleep, brain scans have revealed brief but powerful bursts of neuronal firing known as sleep spindles, predominantly seen in non-REM sleep stage N2. These spindles are associated with synaptic plasticity and play a crucial role in memory consolidation. Memory consolidation occurs primarily during deep sleep, involving both REM and slow-wave sleep. In SWS, the hippocampus actively replays daily events, stimulating synapses to strengthen neural connections, thereby facilitating the transfer of memories from short-term to long-term storage. Motor skills are also consolidated during this phase. During REM sleep, the amygdala helps to consolidate emotional memories, highlighting the importance of this phase in processing emotions associated with memories.

Getting adequate deep sleep is crucial for the brain to solidify daily memories and learning. Insufficient SWS can impair the ability to retain information and learn new things. However, it is the cycling through both SWS and REM sleep phases that is most vital for proper memory formation and retention.

Sleep deprivation

Sleep is a significant and essential biological requirement for all mammals. Sleep deprivation is defined as either a complete lack of sleep or a shorter than optimal sleep time. Horses that are unable to lie down are unable to achieve REM sleep and are therefore likely to experience sleep deprivation. Sleep deprivation in the horse can be manifested as excessive daytime sleepiness and collapsing episodes. There are many factors that can impact on the horse's ability to go through all the stages of sleep required for normal healthy functioning:

- environmental stress in the form of excessive noise or temperature
- pain or discomfort
- difficulty getting up and lying down, prevented by, for example, the behaviour of other horses, a stable that's too small, environmental stimuli (noise, events) causing fear or anxiety
- inadequate bedding for those horses stabled at night
- disruption in the herd for grass-kept horses
- being transported long distances
- feeling unsafe, for example a lone horse without a companion to "stand guard"

I have seen a number of behaviour cases of sleep deprivation in horses, usually reported to me at the stage where the horse is collapsing frequently. This is often thought to be narcolepsy, but this is rare in horses (Bertone, 2015; Haines, 2022). In the cases I have seen, a range of the aforementioned factors were involved but the common factor in all of them was undiagnosed pain, making it too uncomfortable or mechanically impossible for the horse to lie down and enter the deeper phases of sleep.

REM sleep has a significant effect on the learning centres of the brain, serving as a critical period for consolidating emotional and procedural memories. Brain scans reveal that activity levels during REM sleep are similar to those during wakefulness, facilitating complex memory processing. This is essential for horses, as insufficient REM sleep can impair their ability to perform learned skills effectively.

Additionally, sleep is vital for maintaining overall health and balance in the body. During sleep, the body secretes hormones crucial for growth and repair. Disrupted sleep can significantly impact these hormone levels in horses, leading to:

- a decrease in growth hormone, which is important for muscle growth and repair, bone strength and coat health

- elevated cortisol levels, which can disrupt immune system functioning
- affected glucose metabolism, increasing the risk of insulin resistance and potential metabolic disorders such as laminitis or equine metabolic syndrome (EMS).
- imbalances in reproductive hormones (testosterone, oestrogen and progesterone), negatively affecting fertility

Overall, a lack of sufficient deep sleep alters hormone secretions that are vital for tissue repair, growth, metabolism, stress response, immunity, glucose regulation and reproductive health.

Memory and its role in behaviour

Learning can be defined as the acquisition of new knowledge, whereas memory is the retention of that learned information. Different parts of the brain, and thus different neural circuits, are involved in the acquisition and retention of learning. The process of memory formation starts when sensory information enters the brain's short-term memory, where the hippocampus begins to encode this information for long-term storage. During sleep, the neural connections specific to that memory are strengthened and consolidated into long-term memories, with the cortex serving as the primary site for long-term storage.

Long-term memory is believed to have virtually unlimited storage capacity, with memories consolidated and permanently stored across neural networks in the cortex. In contrast, short-term memory retains information for up to about 30 seconds and has a very limited capacity. Working memory, which is often discussed in conjunction with short-term memory, involves the active manipulation of stored information to make sense of, modify and interpret it, whereas short-term memory refers more to the simple retention of information over a brief period. Understanding the different viewpoints on working and short-term memory is crucial because there are various theories about how memory functions, reflecting different researchers' perspectives on these cognitive processes.

Working memory "slots"

When I was studying psychology as an undergraduate, we were taught about the cognitive psychologist George A. Miller's idea of a limited number of "slots" in short-term memory. His research highlighted that the average number of objects an individual could hold in working memory was between five and nine, hence the idea of seven memory slots (+/–2). You may hear some psychologists or behaviourists using the phrase "all his/her slots are full" with reference to the inability of an individual (human or equine) to take in any more information, with the idea that if the brain has to take in more information then something has to give way; either the information in one of the "slots" has to be dropped and the information it held is forgotten, replaced by the new piece of information, or alternatively the new information cannot be consolidated.

Later research expanded and nuanced our understanding of short-term and working memory, and the "seven slots model" may be considered somewhat simplistic by today's standards. Memory capacity isn't just about slots but also involves the richness of information, the individual's cognitive load and how information is processed and encoded; even so, Miller's theory has been supported by numerous experimental studies, and the concept that short-term memory has a limited capacity that can be optimised through strategies like chunking remains valid (Baddeley, 1994; Mathy and Feldman, 2012).

Resource model

When we think about the development of memory as an evolutionary survival strategy it makes sense that paying attention to limited but essential information whilst also excluding non-life-threatening information is important. The prey animal must focus on the hostile environment in order to recognise a possible attack by a predator. Considering memory in this light it seems obvious that there must be a mechanism whereby the increasing salience or relevance of the information/stimuli cause it to take priority in both short-term recognition and long-term storage. Rather than being limited to a short-term fixed storage solution (the seven slots), a growing body of evidence indicates that memory resources can be unevenly distributed so that prioritised items are stored with enhanced precision. This comes with a corresponding cost to other items which are recalled with less precision. What this suggests is that the allocation of limited memory (whether working or short term) is likely to be controlled and updated with changing behavioural priorities (Ma *et al.*, 2014).

Types of memory

Memory can be categorised into declarative and non-declarative, or procedural, memory. Declarative memories are those that are explicitly available to be recalled such as information (where the water trough is), events or specific stimuli. They are often easy to form and relatively easy to forget, although relative salience is a key factor here. Procedural memories, on the other hand, include instances of perceptual, stimulus-response and motor learning of which we (or our horses) are not consciously aware. The human example that always seems to be offered is riding a bike. The equine example I usually use is learning to do shoulder in. Procedural memories operate automatically and relate to skills and behaviours, which can be performed without conscious recollection (Cowan, 2008).

Memory retrieval

While horses have a more limited prefrontal cortex, their memory systems share notable similarities to humans. In general, memories are retrieved in three ways – recall, recognition and context cues. Recall involves the conscious retrieval of information, where horses access memories through context, familiar stimuli or reward motivation. Recognition is the identification of previously learned information when re-exposed to it through familiarity, and it is an easier form of retrieval than recall. Context cues aid retrieval from long-term storage, and this type of memory retrieval is particularly strong in the horse due to the amygdala's involvement in encoding memories with emotional arousal, playing a significant role in how horses respond to their environment.

The differences between short-term and long-term memory play an important role in behaviour and training. Short-term memory enables the horse to respond to immediate stimuli like sights, sounds and cues; essential for quick behavioural responses and reactions and making it ideal for use in the early stages of training basic associations. However, these responses are easily disrupted by distractions or gaps between cues. Long-term memory allows the horse to store information for years and enables complex skill and task learning over time as well as generalising behaviours across contexts. This is what enables us to back a youngster in late summer one year, turn them away all winter and have them come into work the following spring having remembered their earlier rudimentary training experience. Memory consolidation relies on practice and repetition for robust encoding with learning responses becoming habitual through long-term storage. For successful training, short-term memory helps horses to link cues, while ingraining skills in long-term memory results in reliable learned behaviours under stimulus control.

The interaction between health and behaviour

Ailments, diseases and physiological imbalances have the potential to cause discomfort, pain or distress that horses then communicate to us and other horses through their behaviour. Recognising behavioural changes as potential indicators of health issues and addressing those underlying health issues is a key element in preventing and resolving many behavioural problems in horses.

Pain can show up as the horse becoming irritable, aggressive, reluctant to move or reactive to touch. Digestive issues are common in horses and can lead to changes in appetite and food-related behaviours. Dental problems can manifest as horses resisting the bit or having trouble eating. Intestinal worms and other parasites might be responsible for itching and abdominal pain leading to altered behaviours, and endocrine imbalances such as Cushing's disease cause hormonal shifts that can markedly affect temperament and activity level. Musculoskeletal issues are common given the nature of the physical activities that we expect of horses, under the load of a rider, often at speed or requiring complex movements of the body in dressage for example.

Long-term stress, including stress as a result of pain or following pain, creates chronic arousal of the autonomic nervous system and activation of the HPA axis. The release of stress hormones increases susceptibility to disease and dysfunction, typically causing inflammation and immunosuppression which can manifest as sweet itch and other skin conditions, susceptibility to injury and other types of disease and dysfunction.

Identifying pain through behaviour

There are a number of scientific studies that have looked at the link between pain and behaviour in horses and endeavoured to provide tools for the recognition of pain/discomfort which may be contributing to changes in a horse's behaviour:

- The Horse Grimace Scale (Dalla Costa *et al.*, 2014) was developed as a pain assessment tool during routine castration. It takes six areas of the face and scores them based on whether there are signs of discomfort – not present (0), moderately present (1) or obviously present (2). Since then the scale has been tested in other situations such as with horses undergoing dental treatment (Coneglian *et al.*, 2020). The scale is now being used to assess degrees of pain in horses, and once you get your eye in and become familiar with recognising the facial indicators it can be very useful. Although it's not a definitive measure of discomfort, as a recent study showed, participants were unable to distinguish between horses diagnosed with stomach ulcers and healthy individuals with the use of the grimace scale alone (Ferlini Agne *et al.*, 2023).
- Gleerup's Pain Face (Gleerup *et al.*, 2015) identified the following indicators: "low" and/or "asymmetrical" ears, an angled appearance of the eyes, a withdrawn and/or tense stare, nostrils widened towards the outside, and tension of the lips, chin and certain facial muscles to indicate the presence of pain/discomfort.
- The ridden horse pain ethogram (RHpE) lists 24 behavioural characteristics that can be displayed by horses when ridden and that correlate to pain and subclinical lameness (Dyson, 2022). The majority of these characteristics are at least ten times more likely to be seen in lame horses compared with non-lame horses. The observation of more than eight out of the 24 behaviours is likely to reflect the presence of musculoskeletal pain.

- The equine discomfort ethogram (non-ridden) is based on 35 years of research work in an equine hospital comparing the behaviour of both normal and physically uncomfortable horses. The researchers catalogued over 60 behaviours associated with equine physical discomfort and grouped them into eight categories: posture and weight bearing; limb and body movements; head, neck, mouth and lip movements; attention to area; ear and tail movements; overall demeanour; altered eating or drinking and vocalisations/audible sounds (Torcivia and McDonnell, 2021).

For those of us with older equines it might be worth considering drawing up an older horse ethogram, listing the types of behaviours that would indicate a good or satisfactory quality of life for that individual older horse. I did this for my horse when I knew that I should measure his time left with me in months, not years; I designed a daily tick sheet with the following behaviours:

Eating grass
Eating hay
Browsing
Using a treat ball
Drinking
Walking
Trotting
Rolling
Scratching on tree or fence
Mutual grooming with companion
Playing with companion

There was an element of weighting in these elements, of course; not seeing playing for two or three days was not as significant as not seeing walking would have been! Cantering would have been a bonus that didn't need to be listed.

This proved to be a useful and (partially) objective tool to assess potential pain/discomfort in my older horse (age 32); I had some months of data which helped me in discussions with my vet and helped me to make the inevitable and emotional decision when the time came.

If you are ever in a position to do the same with your elderly horse, one shortcoming that I had reckoned without (despite the detailed discussion of conditioned stimuli in earlier chapters!) was how quickly Rusty learned to know when I was about, "spying" on him. First he recognised me, of course, then he recognised my car when I tried to hide in that, then the sound of the car engine when I tried to park further away, then my shadowy presence hiding behind a hedge . . . something to think about in advance, nowadays an outdoor camera would be a helpful addition to this toolkit!

References

Aleman, M., Williams D. C., & Holliday, T. (2008). Sleep and sleep disorders in horses. *AAEP Proceedings*, *54*.

Baddeley, A. (1994). The magical number seven: Still magic after all these years? *Psychological Review*, *101*(2), 353–356.

Bertone, J. J. (2015). 10 sleep and sleep disorders in horses. *Equine Neurology*, *123*.

Coneglian, M. M., Borges, T. D., Weber, S. H., Bertagnon, H. G., & Michelotto, P. V. (2020). Use of the horse grimace scale to identify and quantify pain due to dental disorders in horses. *Applied Animal Behaviour Science*, *225*.

Cowan, N. (2008). What are the differences between long-term, short-term and working memory? *Progress in Brain Research*, *169*, 323–338.

Dalla Costa, E., Minero, M., Lebelt, D., Stucke, D., Canali, E., & Leach, M. C. (2014). Development of the Horse Grimace Scale (HGS) as a pain assessment tool in horses undergoing routine castration. *PLoS One*, *19*, 9.

Dyson, S. (2022). The ridden horse pain ethogram. *Equine Veterinary Education*, *34*, 372–380.

Ferlini Agne, G., May, B. E., Lovett, A., Simon, O., Steel, C., Santos, L., Barbosa, B., Werner, L., Daros, R., Somogyi, A., Sykes, B., & Franklin, S. (2023). Horse grimace scale does not detect pain in horses with equine gastric ulcer syndrome. *Animals*, *13*(10), 1623.

Gleerup, K. B., Forkman, B., Lindegaard, C., & Andersen, P. H. (2015). An equine pain face. *Veterinary Anaesthesia and Analgesia*, *42*(1), 103–114.

Greening, L., & McBride, S. (2022). A review of equine sleep: Implications for equine welfare. *Frontiers in Veterinary Science*, *9*, 916737.

Haines, A. (2022). *Sleep Deprivation in Horses*. Available at Veterinary-practice.com/article/sleep-deprivation-in-horses (Accessed 28 April 2024).

Hebb, D. O. (1949). The first stage of perception: Growth of the assembly. *The Organization of Behavior*, *4*(60), 78–60.

Hemmings, A. (2023). *Training the Brain: A Head First Journey into the Neuroscience of Learning*. Available at https://equinebehaviouraffiliation.org/beyond-behaviourism-webinar-series/ (Accessed October 2023).

Henshall, C., Randle, H., Francis, N., & Freire, R. (2022). The effect of stress and exercise on the learning performance of horses. *Scientific Reports*, *12*(1), 1918.

LeDoux, J. E. (1998). *The Emotional Brain: The Mysterious Underpinnings of Emotional Life*. Simon and Schuster.

Ma, W. J., Husain, M., & Bays, P. M. (2014). Changing concepts of working memory. *Nature Neuroscience*, *17*(3), 347–356.

Mathy, F., & Feldman, J. (2012). What's magic about magic numbers? Chunking and data compression in short-term memory. *Cognition*, *122*(3), 346–362.

Patel, A. K., Reddy, V., Shumway, K. R., & Araujo, J. F. (2022). Physiology, sleep stages. StatPearls. *StatPearls Publishing. Viitattu*, *3*, 2023.

Tiwari, R., Kumar, R., Malik, S., Raj, T., & Kumar, P. (2021). Analysis of heart rate variability and implication of different factors on heart rate variability. *Current Cardiology Reviews*, *17*(5).

Toates, F. M. (2007). *Biological Psychology*. Pearson Education.

Torcivia, C., & McDonnell, S. (2021). Equine discomfort ethogram, *Animals*, *11*(2).

Part 3

Equine welfare

Chapter 7

Welfare and wellbeing

Defining equine welfare and wellbeing

Equine welfare has been referred to as the basic standards of care needed to meet a horse's physical needs: providing proper nutrition, water, shelter, health care, safety and freedom from abuse, neglect or unnecessary discomfort. The assessment of welfare standards usually involves looking at resource provision and management factors. Equine wellbeing is typically thought of as a broader concept that encompasses welfare but also focuses on optimising a horse's overall quality of life. Wellbeing factors include positive mental and emotional experiences, good social human-horse relationships, the ability to display natural behaviours, adequate exercise and enrichment and low-stress handling. Wellbeing implies reaching an ideal state where the horse not only has their basic needs met but also experiences companionship, pleasure, interest and happiness. Many researchers now talk about the distinction between a "life worth living", meaning a minimum acceptable standard of living, and "a good life", as an optimum level of experience.

The most reliable indicators of wellbeing are good health and the manifestation of a normal behavioural repertoire. Studies of free-roaming horses are useful as indicators of behaviour that are important to horses in general. Researchers who can study individual managed horses, how they experience their lifeworld and their individual emotional state, are increasingly focusing on providing horses with options, allowing them to choose between different environments and conditions and measuring the strength of their preference. In this sense we are starting to become more aware of the horse's own "agency" in terms of the way their lives are lived. As an example, in a 2019 project researchers decided to "ask" a group of 23 horses (ten who usually wore rugs and 13 who didn't) what they preferred based on the current weather conditions (Mejdell *et al.*, 2019). The horses were trained to choose between three symbols representing:

1. Rug on
2. Rug Off
3. No change

They were tested under conditions where they had been outdoors in their paddocks for two hours. It took two weeks for the scientists to feel confident that the horses understood this method of communication and were able to consistently state their preferences. The results showed that the horses had a clear preference for wearing a rug when temperatures were low,

DOI: 10.1201/9781003581390-10

Figure 7.1 The insulating effect of a horse's full coat. Photograph by Rita Kochmarjova.

in this case below –10°C, whilst when air temperatures were above 10°C very few of the horses chose to wear a rug. When temperatures were moderate (5°C–10°C) but it was wet and windy, the horses showed a clear preference for having a rug on, which matches the traditional horse carer's view that wet and windy weather is the most disliked type of condition, when horses will be more likely to seek shelter. You might also expect a similar result when it was snowing, but the researchers actually found that fewer horses preferred to wear a rug in these conditions. They speculated this was likely because the group of horses were well adapted to living in cold weather, being from Norway and Sweden, and that snow tends to sit on the top of their (full) coats rather than penetrating the insulating layer of hair (Figure 7.1). What this study and others are demonstrating is that the decisions we make for our horses may not be the same decisions our horses would choose for themselves.

When we look at how we can enhance the experience of our managed horses a crucial element of their day to day experience is the extent to which they can exercise choice and control over their environment, and this has a significant impact on their wellbeing. In a free-roaming horse's day they are almost constantly making what appear from a human perspective to be "decisions", or at least taking initiating action, about what they will do next. An inability to make these choices, either because the horse's space is restricted or because the resources to engage in that behaviour are not present, can contribute to stress and anxiety and impact horses' facility to experience "a good life". Animals in captivity, including domestic horses, have the vast majority of choices made for them by their caretakers. They also have little or no control over the environment surrounding them. However, researchers have found that even the smallest opportunity to make choices can have a positive effect on behaviour (Boissy *et al.*, 2007). In practice on a stable yard this could even be as little as having two different water containers, perhaps a bucket and an automatic dispenser, allowing horses to exercise choice over which one to drink from.

How can welfare and wellbeing be measured?

Those of us involved with horses will undoubtedly be familiar with the main checks we can and should be doing to assess whether our horse's basic needs are being met:

- Coat condition – the horse's coat is shiny and smooth, indicating good nutrition and overall health.
- Hoof condition – hooves are properly trimmed and shaped.
- Nutrition assessment – dietary needs (feeding and forage) are being met.
- Physical health – checking for injury, disease, lameness, dental issues and parasites as well as measurements such as temperature and heart rate which can help reveal the health status of the horse. Triangulating physical and behavioural observations with cortisol levels or heart rate variability provides information on stress status.
- Body condition scoring – assessing the horse's physical condition to determine their level of risk based on being over or under their ideal weight.
- Environment – evaluating the horse's living space for safety, cleanliness, size, social contact and enrichment. The environment in which the horse is living can have a significant impact on their welfare.
- Behavioural observations – looking out for signs of contentment (relaxed, active, playful, attentive) or discontent (aggression, irritability, depression) and note incidences of repetitive behaviours such as box walking, crib biting and wind sucking. Abnormal behaviours may indicate poor welfare.
- Training/handling assessment – evaluating if training and handling methods are based on the scientific principles of learning theory and seek to minimise the use of aversive practices. Harsh methods negatively affect welfare. Least intrusive, minimally aversive (LIMA) principles have until recently been the choice of many; however, recently Fernandez (2024) has proposed a different model to guide ethical training practices: least inhibitive, functionally effective (LIFE), which provides a helpful empirical framework and facilitates a clearer connection between training practices and behavioural and welfare sciences.

This combination of physical examinations, behavioural observations and environmental review provides a thorough assessment of equine welfare and can identify areas for improvement. But when we start to consider measuring our horse's overall wellbeing we need to go further by:

- focusing on positive wellbeing indicators, not just the absence of negative states. We do this by assessing the horse's overall quality of life and positive affective states.
- evaluating the strength of social bonds between horses as an indicator of wellbeing including positive social behaviours like mutual grooming.
- assessing the human-horse relationship and interactions. A strong bond based on positive communication and relationships reflects enhanced wellbeing.
- considering the horse's level of cognition and learning ability as a potential indicator of inner state. Horses with positive emotions may be better/quicker learners.
- analysing time budgets and natural behaviours. Horses should display normal, species-typical behavioural repertoires which provides assurance that behavioural needs are being met.
- providing environments and management that allow horses to express natural behaviours like near-constant grazing/feeding, social contact and ample space.
- developing validated quality of life assessment tools for horses that incorporate both physical and psychological wellbeing factors.
- focusing assessments at the individual level in addition to herd/population measures in order to assess the subjective wellbeing of each individual horse.

Overall, a combination of factors needs to be assessed including social, psychological, physical and environmental with a focus on delivering positive experiences and affective states in order to gauge equine wellbeing at both a group and an individual level (Kelly *et al.*, 2021; Lesimple, 2020).

Identifying good and poor welfare and wellbeing

Good equine welfare meets animals' basic health and behavioural needs and minimises negative mental states. Poor welfare fails to address physical problems or psychological distress, subjecting horses to unsafe or restrictive conditions that prevent natural behaviour and cause suffering (Table 7.1). Keep the following tables in mind when you read further on in this chapter about the five domains of animal welfare model and the five provisions that emerged from this model.

Going further to consider wellbeing, good wellbeing attends to both the physical and mental/emotional factors that lead to positive experiences for horses. It is not just concerned with meeting their basic needs and ensuring they are free from suffering. Poor wellbeing is indicated by behaviours showing psychological distress as well as unaddressed physical issues and an environment unsuited to the horse's needs (Table 7.2) (Fraser, 1997).

It may seem on the face of it that the factors that indicate poor welfare and those that indicate poor wellbeing are remarkably similar; however, the contrast between good welfare and good wellbeing is significant. Indeed in her book *Equine Education*, Marthe Kiley-Worthington (2004) asserts that horses also have intellectual needs including opportunities to use their brains, make decisions and choices as well as to acquire knowledge by having a life full of different experiences and situations to adapt to, something to keep in mind when we consider the concept of agency under the Five domains framework, discussed later.

Table 7.1 Features of Good and Poor Equine Welfare

Good equine welfare	Poor equine welfare
Access to clean water Balanced diet Unrestricted feeding	Lack of food/water Unbalanced diet Restricted feeding
Protection from injury/disease Veterinary care as needed Hoof care as needed Dental care as needed	Lack of veterinary care Repeated or untreated injuries imply poor conditions Signs of abuse/neglect
Shelter Well-maintained facilities	Small/unclean living space
Social contact Free movement Opportunity to graze	Social isolation Confined movement Performance of abnormal behaviour – stereotypies, aggression
Low-stress training methods Minimal restraint	Excessive restraint Aversive, highly stressful training methods
No signs of boredom, frustration, fear or distress	Signs of apathy/depression, fear, frustration and distress

Table 7.2 Features of Good and Poor Wellbeing

Good wellbeing	Poor wellbeing
Horse appears content, relaxed, attentive and engaged, indicating a positive mental state Opportunity to express a range of emotions including joy, happiness, anger, sadness	Signs of boredom, fear, frustration, apathy Restricted ability to express the full range of emotions
Free from injury, disease and illness Receive regular veterinary, dental and hoof care to maintain health Unrestricted feeding/choice of food	Repeated/untreated lameness/illness Poor nutrition/restricted feeding
Strong horse-human bond Trusting relationships with human caregivers Effective communication between horse and human focused on understanding Opportunity to form long-term bonds/friendships with other horses	Weak horse-human bond Lack of trust between horse and human Ineffective cues Punishment No opportunity to develop long-term bonds/friendships
Opportunity to demonstrate natural behaviours – grazing, exploring, socialising with other horses Freedom to choose when and where to eat, sleep, play	Confined Isolated Prevented from socialising How they spend their time dictated by yard routine/caregivers, etc.
Stimulation – opportunity to learn about their environment and social relationships Social play and enrichment	Unchanging routine Lack of stimulation/variety/enrichment
Low-stress handling – patient training and positive reinforcement	High stress/harsh handling – coercive training methods
Open, airy, spacious conditions Ample opportunity to move, with time budgets similar to free-roaming horses	Small crowded conditions Lack of pasture time Restricted movement/enforced stillness, e.g. standing in stables or small paddocks

Why is it important to understand wellbeing and welfare in the context of equine behaviour?

Horses are prey animals and their natural behaviours and needs stem from being a prey species – wanting to be part of a herd and needing open spaces and movement. This places them in rather a fish out of water situation when we expect them to live domestically in individual stables, hack out alone, enter dark, enclosed trailers and change home range at times not of their choosing. To a degree, horses are good at habituating to unfamiliar experiences, but when we take too much advantage of that, behaviour difficulties can emerge. Providing for horses' innate behaviours promotes better welfare. Equally, the horse's behaviour is a window into their mental state. If we accept the premise that wellbeing relates to the horse's overall mental/emotional experiences, then observing behaviour is one of the main ways we can assess their level of wellbeing. The expression of normal, natural behaviour generally indicates good or positive wellbeing. Stereotypic behaviour, lethargy, aggression and other abnormal behaviours may signal that the horse's welfare is being compromised. They suggest a mismatch between the horse's needs and their environment/management. In addition, any changes in a horse's normal behaviour may indicate the presence of physical pain or mental distress.

Equine ethology research as discussed in chapter 2 provides insight into a horse's behavioural needs (grazing, socialising, moving) which are important for the wellbeing of our domesticated horses. Providing for natural behaviour is a key part of promoting positive wellbeing states, not just avoiding negative ones. Environments and handling that allow horses to express their natural behaviour are likely to enhance their overall wellbeing, and since we cannot directly ask a horse about their subjective experience, behaviour may be one of our best indicators of their internal mental state and therefore their overall wellbeing.

Animal welfare legislation in the UK

The Animal Welfare Act 2006 is the principal law relating to animal welfare in the UK. The act requires owners and keepers to ensure that they meet the needs of any horse, pony, donkey or mule which they are responsible for, whether on a permanent or temporary basis. This includes a duty to look after their basic welfare by:

- providing a suitable environment and place to live
- providing a suitable diet including fresh, clean water
- making sure they are able to behave normally and naturally
- providing appropriate company
- protecting them from pain, suffering, injury and disease
- keeping records of all medical treatment
- keeping up to date with a vaccination programme (under the control and advice of a vet)
- registering them with a vet practice and a registered farrier
- ensuring they have a valid passport and microchip

The act also prohibits animal cruelty, which includes causing unnecessary suffering to an animal. Enforcement of equine welfare offences can result in fines, imprisonment, seizure of animals and disqualification from keeping animals in the future (Gov.UK, 2023).

Other key UK laws include the Animal Health and Welfare (Scotland) Act 2006 covering welfare in Scotland, and the Welfare of Animals Act (Northern Ireland) 2011, covering welfare in Northern Ireland. The Animal Welfare (Sentencing) Act 2021 increased maximum prison sentences for animal cruelty offences in England and Wales, and the Animal Welfare (Sentience) Act 2022 recognises the sentience and welfare needs of all vertebrate animals, requiring policy-makers to consider animal sentience when creating or changing laws, and established the Animal Sentience Committee to examine and review government policy with reference to effects on animal welfare. Statutory regulations surrounding specific issues like transport, slaughter and research animals also contain welfare provisions. Enforcement is generally carried out by local authorities, animal health/welfare inspectors and police, and the policy overview involves both governmental and independent advisory bodies.

Models of animal and equine welfare

There are several models of animal welfare, and we start with one of the historical pillars, Brambell's Five Freedoms of Animal Welfare model. Developed in 1965 in response to the public outcry in the UK to Ruth Harrison's 1964 book *Animal Machines* and following a government-commissioned investigation into livestock production, the Brambell report significantly influenced animal welfare standards, including those pertaining to equines.

Brambell's five freedoms of animal welfare:

1. Freedom from hunger and thirst
2. Freedom from discomfort
3. Freedom from pain, injury or disease
4. Freedom to express normal behaviour
5. Freedom from fear and distress

The five domains model

Next we move on to what is perhaps currently the most widely known welfare model in equine circles, the five domains model (Figure 7.2). This was first proposed by Professor David Mellor and Dr Cam Reid in 1994 in the context of laboratory animal welfare and has been developed regularly since then to incorporate the latest research in animal welfare science. In their 1994 paper (Mellor and Reid, 1994) the authors proposed that good welfare or wellbeing is a state of being, manifest in an animal when their nutritional, environmental, health, behavioural and mental needs are met. They related these needs back to the five freedoms, which they used as a basis for developing a new system of assessing the impact on the animal of being used in laboratory experiments, and in this way they transformed the freedoms into "domains of potential compromise"; thus, they redefined them to emphasise the *extent of welfare compromise* rather than the earlier ideal of the absence of compromise.

The five domains

1. Nutritional conditions – this domain considers the animal's nutritional status and any imbalances, deficiencies, or excesses that may compromise welfare. Factors examined include access to food and water, diet and incidence of malnutrition.
2. Physical environmental conditions – this domain covers the impact of living conditions. It considers issues like shelter, space, temperature, air quality, hygiene and noise. The aim is to identify environmental conditions that may pose challenges the horse cannot control.
3. Health conditions – this domain considers health status, disease, injury and levels of physical fitness. The focus is on identifying health conditions and their associated affects both positive and negative.
4. Behavioural interactions – this domain considers the concept of agency when the horse interacts with their environment, other non-humans and human beings. Agency refers to an animal's ability to actively engage in their environment and interactions in order to gather knowledge, enhance their skills and make conscious choices to behave in a particular way (Littlewood, 2023).
5. Mental state – this domain integrates the results from the other four domains to make inferences about the animal's likely affective experiences and therefore provide an assessment of the overall welfare state of the individual horse.

(Mellor *et al.*, 2020)

The five domains model provides a systematic framework for assessing welfare in a scientifically grounded way across five key domains, including attending to animals' positive affective experiences. It is now recognised that what matters to horses (and all animals) in welfare terms is their subjective experience. In essence, there is an acceptance that "a life worth living" cannot be achieved simply by mitigating or avoiding negative experience; animals need pleasurable experiences as well.

The Five Domains Model

Physical/Functional Domains

Survival-Related Factors						Situation-Related Factors	
1: Nutrition		**2: Environment**		**3: Health**		**4: Behaviour**	
Restrictions on:	*Opportunities to:*	*Unavoidable/imposed conditions:*	*Available conditions:*	*Presence of:*	*Little or no:*	*Exercise of 'agency' impeded by:*	*'Agency' exercised via:*
Water intake Food intake Food quality Food variety	Drink enough water Eat enough food Eat a balanced diet Eat a variety of foods	Thermal extremes Unsuitable substrate Close confinement Atmospheric pollutants: CO_2, ammonia, dust, smoke	Thermally tolerable Suitable substrate Space for freer movement Fresh air	Disease: acute, chronic Injury: acute, chronic; husbandry mutilations Functional impairment: due to limb amputation; or lung, heart, vascular, kidney, neural or other problems	Disease Injury Functional impairment	Invariant, barren environment (ambient, physical, biotic) Inescapable sensory impositions Choices markedly restricted	Varied, novel, engaging environmental challenges Congenial sensory inputs Available engaging choices
Voluntary overeating Force-feeding	Eating correct quantities	Unpleasant/strong odours Light: inappropriate intensity Loud/otherwise unpleasant noise	Pleasant/tolerable odours Light intensity tolerable Noise exposure acceptable	Poisons	Poisoning	Constraints on environment-focused activity Constraints on animal-to-animal interactive activity	Free movement Exploration Foraging/hunting Bonding/reaffirming bonds
		Environmental monotony: ambient, physical, lighting	Normal environmental variability	Obesity/leanness	Body condition appropriate		Rearing young
		Unpredictable events	Predictability	Poor physical fitness: muscle de-conditioning	Good fitness level	Limits on threat avoidance, escape or defensive activity Limitations on sleep/rest	Playing Sexual activity Using refuges, retreat, or defensive attack Sleep/rest sufficient

Affective Experience Domain

5: Mental State

Negative	*Positive*	*Negative*	*Positive*	*Negative*	*Positive*	*Negative*	*Positive*
Thirst Hunger (general) Hunger (salt) Malnutrition malaise	Wetting/quenching pleasures of drinking Pleasures of different tastes/smells/textures Pleasure of salt taste Masticatory pleasures Postprandial satiety	*Forms of discomfort:* Thermal: chilling, overheating Physical: joint pain, skin irritation Physical: stiffness, muscle tension Respiratory: e.g. breathlessness Olfactory Auditory: impairment, pain Visual: glare/darkness eye strain	*Forms of comfort:* Thermal Physical Respiratory Olfactory Auditory Visual	Breathlessness Pain: many types Debility, weakness Sickness, malaise Nausea Dizziness	Comfort of good health and high functional capacity	Anger, frustration Boredom, helplessness Loneliness, isolation Depression Sexual frustration	Calmness Engaged, in control Affectionate sociability Maternally rewarded Excitation/playfulness Sexual gratification
Bloated, over full Gastrointestinal pain	Gastrointestinal comfort	Malaise from unnatural constancy	Variety-related comfort	Physical exhaustion	Vitality of fitness	Anxiety, fearfulness, panic, anger Neophobia Exhaustion	Secure/protected/confident Likes novelty Energised/refreshed

Welfare Status

Figure 7.2 The five domains model (modified from Mellor and Beausoleil, 2015): the examples provided for the physical/functional Domains 1 to 3, labelled "Nutrition", "Environment" and "Health", are intended to direct attention towards mainly *internal* survival-related factors, and those provided for Domain 4, labelled "Behaviour", focus attention largely on *external* situation-related factors. For each of Domains 1 to 4, examples of negative and positive factors are provided and are aligned with inferred negative or positive affective experiences, assigned to Domain 5, labelled "Mental State". The overall affective experience in the mental domain equates to the welfare status of the animals, as explained in the text. Note that an animal exercises "agency" (Domain 4: "Behaviour") when it engages in voluntary, self-generated and goal-directed behaviours. Reproduced with permission (Mellor, 2017).

Rather than defining good and bad welfare, the model describes areas that are relevant to welfare assessments based on the scientifically supported understanding that horses may have pleasurable experiences when their external circumstances include:

- variability – providing a balance between predictability/controllability and novelty/unpredictability
- meeting species-specific needs for movement and exercise
- access to preferred sites for resting, thermal comfort and elimination behaviours
- environmental choices that encourage exploratory and foraging behaviours
- availability of a variety of feeds having attractive smells, tastes and textures
- circumstances that enable social species to engage in bonding activities with familiar members of the same species (conspecifics)

(Mellor *et al.*, 2020)

The five domains model enables us to provide examples of positive and negative circumstances which may impact the horse's (or any animal's) affective state. By identifying areas of imbalance or disruption we can highlight where there may be an increased risk of negative affect and the potential for welfare problems to arise (McGreevy *et al.*, 2018).

For readers interested in learning more about the application of the five domains model, I recommend taking the online course "Applying the Five Domains Model to the Welfare Assessment of Sport and Recreation Horses" developed by Cristina Wilkins (2023) and hosted by the University of New England.

The five provisions

Alongside the five domains, Mellor (2016) and colleagues developed the concept of the five provisions, listed in Table 7.3 and emphasising good nutrition, environment and health, suitable conditions to exhibit appropriate species-specific behaviour and positive mental experiences including, crucially, a sense of control over their environment and actions within it.

Table 7.3 The Five Provisions of Good Welfare

Provision	Welfare Aims
1. Good nutrition: provide ready access to fresh water and a diet to maintain full health and vigour	Minimise thirst and hunger and enable eating to be a pleasurable experience
2. Good environment: provide shade/shelter or suitable housing, good air quality and comfortable resting areas	Minimise discomfort and exposure and promote thermal, physical and other comforts
3. Good health: prevent or rapidly diagnose and treat disease and injury, and foster good muscle tone, posture and cardiorespiratory function	Minimise breathlessness, nausea, pain and other aversive experiences and promote the pleasures of robustness, vigour, strength and well-coordinated physical activity
4. Appropriate behaviour: provide sufficient space, proper facilities, congenial company and appropriately varied conditions	Minimise threats and unpleasant restrictions on behaviour and promote engagement in rewarding activities
5. Positive mental experiences: provide safe, congenial and species-appropriate opportunities to have pleasurable experiences	Promote various forms of comfort, pleasure, interest, confidence and a sense of control

Agency in the behaviour domain

Agency: the subjective awareness of initiating, executing and controlling one's own volitional actions in the world. When we look at how we can enhance the experience of a confined or intensively managed horse, a crucial element of any animal's day to day experience is the extent to which they can exercise choice and control over their environment, and this has a significant effect on welfare. In a free-roaming horse's day they are almost constantly making choices about what they will do next, whether it's run round the field for five minutes with their friends, browse on a particularly tasty piece of tree, or stand in just the right spot where the sun warms their back the best. An inability to make these choices, either because the horse's space is restricted or because the resources to engage in that behaviour are not present, will contribute to stress, anxiety and compromised welfare. Animals in captivity, including domestic horses, have the vast majority of choices made for them by their human caretakers. They also have little or no control over the environment surrounding them; it behoves us to think as creatively as we need to, to offer them choices in their lifeworld that will allow them to exercise choice and agency, understanding also how this correlates with the interplay between appetitive (goal-seeking) and consummatory (goal-achieving) behaviours that were discussed in chapter 4.

Adaptive expendability

The five domains model looks at welfare from an individual horse's perspective in terms of their subjective experience. At around the same time that Mellor and Reid were at work on developing their early model, other researchers were considering the links between evolution and behaviour: Barnard and Hurst's (1996) model of adaptive expendability is a concept in animal behaviour that examines the expendability of certain behaviours from an evolutionary perspective. It helps to explain why some behaviours that appear crucial in one context may be less critical or even absent in another, reflecting a flexibility in behavioural strategies that can enhance survival and reproductive success under changing conditions. The significance of this model lies in its application to understanding how animals adapt to varied environments and manage their energy and resources, which is especially relevant in discussions of equine welfare and management, where natural behaviours might be compromised or need to be encouraged depending on the circumstances.

Adaptive expendability refers to the idea that animals may adapt to environmental stressors in ways that incur a cost to the individual's welfare but provide an overall benefit for the species in terms of reproductive success. For instance, social isolation may result in a horse developing a coping strategy such as weaving which allows them to tolerate their environment but at the cost of potential injury due to the repetitive nature of the movement. Coping with their situation (living conditions, stressors, etc.) costs the horse biologically. If the cost and demands become too high they stop being able to cope biologically (Nordenfelt, 2006) and this will impact their ability to reproduce.

An adaptive expendability approach to welfare assessment evaluates the costs and benefits of these coping strategies in terms of impacts on the individual animal versus contributions to reproductive success. It aims to understand the subjective experience of the horse based on these adaptive trade-offs. Costs to welfare may be considered acceptable if overall reproductive success is significantly increased. This approach contrasts with traditional views that emphasise preserving the individual at all costs. The adaptive expendability approach accepts some costs to the individual to maximise the success of the wider horse population. This type of welfare assessment therefore focuses on allowing and evaluating natural or acclimatised adaptive strategies, rather than simply reducing pressures on the individual. An adaptive expendability welfare assessment might consider the effect on growth, longevity, offspring production, immuno-suppression, disease and injury as well as fluctuations in health and behaviour of the herd as a whole (Barnard and Hurst, 1996).

We can apply Barnard and Hurst's adaptive expendability model to consider a number of the current debates in equine welfare including breeding, management and training practices, for example:

Breeding practices: the expendability model can be used to scrutinise the focus on traits that enhance short-term performance (such as extravagant movement in dressage horses) at the potential expense of long-term health. Traits that lead to early reproductive success and fast maturation might be selected for, even if they contribute to joint problems or other health issues later in life. Recognising this can instead encourage breeders to consider long-term health and vitality when selecting traits, thereby improving the overall welfare and lifespan of horses.

End of career management: horses, especially those in competitive disciplines, often face a sudden change in lifestyle and care once they no longer meet the goal-directed needs of their owner. The expendability model can highlight the importance of planning for this stage of a horse's life, ensuring that they are not simply "discarded" but continue to receive proper care and management (Long *et al.*, 2022).

Veterinary care and intervention: understanding that certain traits or conditions in horses might be a result of evolutionary expendability can guide veterinary practices. For instance, certain age-related diseases might not have been "selected against" evolutionarily because they occur post-reproduction. Recognising these patterns can lead to better preventative care strategies that extend the active and healthy lives of horses.

Training and handling: the expendability model can be applied to training methods. It's recognised that young horses are pushed to perform early in their lives (Gilbert and Gillett, 2011), which aligns with an "expendability" view where early life performance is prioritised without as much regard for long-term welfare. A shift towards gentler, long-term focused training can emerge from understanding the detrimental effects of early, intense training.

Resource allocation: In broader terms, this model can inform how resources are allocated to equine care. If younger, competitive horses are seen as more "valuable", older or non-competitive horses might receive less attention or resources. By applying the expendability model, facilities might be encouraged to distribute resources more equitably, ensuring all horses receive adequate care throughout their lives.

Integrating insights from the adaptive expendability model into equine management and welfare strategies emphasises a shift from viewing horses as "expendable" after their prime reproductive or competitive periods to valuing them throughout their entire life cycle, leading to improved welfare outcomes and more humane treatment across all sectors.

Qualitative behaviour assessment

Qualitative behaviour assessment (QBA) is a validated scientific method for assessing animals' emotional states and welfare using observations of their dynamic behavioural expression (Wemelsfelder *et al.*, 2001; Wemelsfelder, 2008). In other words, it involves making qualitative judgements about a horse's overall style of interaction and expression through descriptors like "relaxed", "tense" and "curious". QBA aims to provide more space for the animal's perspective as an emotional being in welfare science with the premise that a horse's feelings are connected to and observable through their whole bodily expression and demeanour. Observers integrate details of posture, context and behaviour to assess the "body language" of the horses they are observing, allowing them to make inferences about emotional states such as calm, anxious and happy. Traditionally, scientists have distrusted qualitative terms about animals as being too subjective or anthropomorphic; however, research has now shown that QBA assessments are scientifically reliable and correlate well with quantitative indicators of welfare. QBA is particularly useful as it captures fluctuations in a horse's state in response to different conditions and reflects the horse as a whole "behaver" rather than the sum total of observed separate behaviours; a dynamic living being whose movements are always meaningful and psychologically expressive. The principle that horses (and all animals) are expressive "behavers" rather than assembled strings of "behaviour" connects with other scientific literature exploring horses' agency or the freedom to choose how they behave. The researchers are trained to measure not only *what* the horses do (walk, sniff, rest, etc.) but also *how* they go about what they do.

In essence, qualitative behaviour assessments look at the whole horse and how they dynamically move around, and experience, their environment. It provides us with a wealth of information about the horse's perception of the world, their affective state and also their level of wellbeing. Horses can move in a way that's relaxed or tense; they can be curious and friendly, hesitant and anxious, lethargic and sluggish or lively and explosive. It is these qualitative descriptors that provide the insight into what the horse is experiencing in that environment at that moment in time. These observations allow us to get closer to understanding how our horses are experiencing the world around them, with clear implications for ourselves as owners, riders, trainers and behaviour therapists.

Behaviour as an indicator of wellbeing

In chapter 2 we discussed the "normal", natural behaviour of the horse by looking at the equine ethogram of free-roaming herds. In this section, in contrast, we consider "abnormal" behaviour. In the past, some of these behaviours were referred to as vices, implying that the horse was to blame, when in reality it is our way of managing and handling horses that creates the difficulties they experience in coping with their managed lives. Stereotypic behaviour is rarely, if ever, seen in natural social and free-roaming populations.

Abnormal behaviours can be categorised as follows:

- Behavioural patterns which are not part of the natural ethogram (e.g. crib biting, wind sucking and weaving).
- Behaviours that are part of the natural ethogram but are exhibited with a different frequency or duration, or in a different context, are typically displacement behaviours (e.g. pawing the ground for a prolonged period when tied up outside the stable) but may become obsessive and repetitive, for example tail rubbing.

Other behaviours often referred to in the prior categories include:

compulsive licking/biting/kicking/stomping – normal behaviour until becomes a repeated and rhythmical action

box/fence walking/pacing/circling – repetitive and/or compulsive locomotive behaviour

head tossing/shaking/nodding – normal behaviour, for example to get rid of insects, but can become a repeated and rhythmical action, sometimes viewed as a coping mechanism

self-mutilation – flank, chest or shoulder biting; indicative of extreme stress

tongue playing – moving and twisting the tongue inside and outside the mouth unrelated to food

yawning – follows a stressful event, or in frustrating contexts

vacuum chewing (chewing although no food present) – follows a stressful event; considered a displacement behaviour during stressful situations and often appears in contexts frustrating for horses

bed eating – domestic horses' attempts to forage

excessive wood chewing/eating – linked to limited roughage availability and restricted ability to forage

coprophagy – ingestion of faeces is normal in foals but linked to limited forage availability in adults

heightened aggression –food-related aggression, social aggression

separation anxiety – frantic locomotive behaviour and vocalisations linked to inability to rejoin herd

Figure 7.3 Despite wearing an anti-cribbing collar, this horse crib bites the fence rail. Photograph by Burloncita.

These behavioural patterns can be regarded as types of adaptation strategy or coping mechanism, providing the horse with temporary relief from stress related to needs not met due to suboptimal management, environment or handling. In welfare terms the concern is that the attempt to cope with their welfare-compromised situation exceeds the horse's capacity for adaptation, and the evidence is mixed as to whether performing them actually helps the horse cope or simply indicates attempts to cope. Regardless, their presence signifies shortfalls in the horse's welfare/wellbeing (McGreevy, 2007). At this point it's worth distinguishing between abnormal behaviour and unwanted behaviour. Unwanted behaviours are those which are in line with the horse's natural ethogram but cause problems for the people who are managing, training and handling them, for example, bolting, shying, bucking and nipping are natural behaviours observed in free-roaming populations but are generally considered unwanted by their human handlers (Hothersall and Casey, 2012). Behaviours unwanted by humans can be motivated by both positive and negative emotions in horses, for example fear, pain or play.

Absence or lack of social contact, roughage and environmental stimuli and insufficient exercise have all been linked to the occurrence of abnormal behaviour. Interestingly, these behaviours don't generally manifest immediately and appear to be the result of a build-up of "stressors" that lay the groundwork for these behaviours. Some horses may have a genetic predisposition for developing stereotypic behaviour, and when this is combined with environmental and management factors (restricted food/movement/social interaction) over time it causes their emergence. Just like humans, individual horses cope with stress differently and have different capabilities in terms of active coping strategies. Our familiar neurotransmitter dopamine (which we met in chapter 6) is a significant factor in influencing these active coping strategies. Research has shown that animals who have been exposed to chronic stress and go on to develop stereotypic behaviour have an up-regulation of the dopamine system (Hemmings *et al.*, 2004).

High dopamine levels have been linked to both impulsive and compulsive behaviours. What is perhaps more interesting is the research conducted on the brains of non-stereotypic horses exposed to chronic stress. Their dopamine system is down-regulated and associated with a state of reduced behavioural output that in humans would be called evidence of clinical behavioural depression. In simple terms this helps to explain why, on a large yard of horses living, managed and handled in exactly the same way, only a percentage of them show stereotypic behaviour. Does this mean they are the only ones experiencing stress? Or is there a cohort who are experiencing stress but cope with that stress in a totally different way by a down-regulation of the dopamine system and subsequent depression of their behavioural output? These are the horses that can be the hardest to identify as their "abnormal" behaviour is categorised by the absence of normal behaviours rather than the performing of stereotypies. How do you differentiate between a horse who is standing in their stable or field with their head down, resting, and a horse in the exact same posture but in a state of behavioural depression? Rochais and colleagues (2016) investigated this very question and found that it was possible to identify behavioural markers including a withdrawn posture and sensory inattention, indicating a depressive-like state, and that it was possible to distinguish these horses from those who were calm and relaxed.

Trauma

Individual trauma arises from an incident, a sequence of events, or circumstances experienced by the individual as physically or emotionally damaging or life-threatening. It has enduring negative impacts on their ability to function and on their physical, mental, social and emotional wellbeing (McMillan, 2019). Highly stressful events that generally fall outside what the individual has learned to be their normal experience or their innate nature as a species have the potential to be traumatic, but not every horse will experience them as traumatic. Trauma can be an initiating factor in the development of abnormal behaviour. Examples of traumatic events include abrupt weaning, sudden commencement of intensive training, drastic management changes (such as the move from herd living to individual stable living), temporary isolation or box rest. These types of negative experiences have particularly far reaching consequences if they happen during the early or developmental periods of a horse's life. For instance, crib biting appears to be linked to weaning time with a high incidence of development around six months of age. In contrast, abnormal locomotory behaviour tends not to start developing until around 56 weeks (McBride and Hemmings, 2009). The other interesting aspect of stereotypies is that they involve a need-related drive that develops in an environment with inadequate opportunities for satisfying that need and, once established, the drive to perform the stereotypic behaviour may become a need in itself. Hence it is so important that with regard to the horse's environment, however limited, we satisfy their ethological needs (sometimes requiring a degree of creative thought!) in order to prevent the emergence of stereotypic behaviour.

Attempts are often made to try to physically prevent abnormal behaviour by the use of anti-cribbing collars (Figure 7.3), anti-weaving bars, removal of favoured crib-biting surfaces and substances that can be painted on surfaces to prevent cribbing. Studies suggest that these preventative measures are stressful to the horse, causing an increase in circulating cortisol levels and a raised heart rate (McBride and Cuddeford, 2001). Preventative measures are of particular concern if we consider that these behaviours may constitute a coping response to a suboptimal environment. They are also unlikely to be successful if the underlying causes of the behaviour are not removed. Research consistently shows that there is an association between abnormal behaviour and two management factors in particular – low/restricted forage and restricted social contact. Although genetic factors are likely to be involved, these behaviours only emerge in deficient environments. Management during a horse's early development (early/abrupt weaning and social isolation) may increase the risks of abnormal behaviour in later life, whilst late, foal-centred weaning and group housing have been shown to have protective effects.

It's also worth referring back to chapter 2 where we discussed time budgets. Time budgets are a way of classifying the proportion of time a horse spends carrying out the different behaviours identified in their ethogram during a 24-hour period. One study on free-roaming Camargue horses found they spent 60% of the day eating and 20% standing. In contrast, horses who were kept stabled and fed restricted forage spent only 15% of their day eating, 65% standing (doing nothing in a barren, confined environment) and more time lying down. Another group of managed horses who were kept in group housing with ad lib hay and straw bedding showed a time budget that most closely resembled the one seen in the free-roaming horses (Kiley-Worthington, 1997).

Equine cognition and wellbeing

The relationship between equine cognition and equine welfare is a growing area of study that has far-reaching implications for the care, training and therapeutic treatment of horses. Understanding equine cognition can help us appreciate not only how horses think and perceive their environment but also how these cognitive processes can influence their overall wellbeing. Equine cognition refers to the mental processes that horses use to acquire, process, retain and act on information from their environment. This includes perception, learning, memory, problem-solving and decision-making. In chapter 3 we discussed how horses possess a wide range of sensory abilities, including acute hearing and a broad field of vision. They can perceive even subtle changes in their environment. They are also highly attuned to the emotional states of other horses and humans, often reading body language and vocal signals with great precision, and we also know that misunderstandings or misinterpretations by humans can lead to stressful situations for horses, thus affecting their wellbeing. We also described how horses are social animals with complex herd dynamics. They recognise individual herd members and understand social networks. Social isolation can severely impact a horse's mental health, while positive social interactions can enhance their wellbeing.

Understanding social cognition is crucial when managing group housing, introducing new horses or addressing aggressive or reactive behaviours. It's no surprise, evolutionarily, that horses have spatial cognition capacities allowing them to form mental maps of their home ranges and the spaces beyond, and to navigate these environments skilfully. Ensuring their living spaces are cognitively suitable and enriching, with clear navigation routes, supports mental wellbeing and reduces stress.

In chapters 4 and 5 we detailed how horses are capable learners, how they can generalise information from one context to another and are skilled at observational learning (learning by watching others). They have also demonstrable capacities for memory, learning and categorisation of objects, which indicates a greater degree of cognitive complexity than traditionally thought. Traditional training methods that do not take into account a horse's cognitive abilities can lead to confusion, stress and learned helplessness. On the other hand, ethical (LIMA/LIFE), evidence-based training approaches that respect the horse's cognition can improve their mental wellbeing. Horses show signs of stress and frustration when faced with problem-solving tasks that are beyond their current capacity, therefore setting training goals within their cognitive means and providing achievable challenges contributes to a positive state of wellbeing (Waran *et al.*, 2007).

In the last chapter on the physiology of behaviour we detailed how horses have a robust long-term memory, especially for tasks or events that have strong emotional components. This memory plays a significant role in their behavioural responses; for example, a horse who has had a traumatic experience may show signs of stress or fear when faced with a similar situation, even years later. Recognising the impact of past experiences on current behaviour is essential for effective training, handling and therapy. In addition, horses have demonstrated problem-solving abilities in various studies, often using trial-and-error methods to achieve a goal. They can make decisions based on their assessment of potential risks and rewards. Restrictive environments that do not allow horses to exercise these skills can lead to frustration and decreased mental wellbeing. Horses also appear to differentiate between individual humans and discriminate between previous positive and negative interactions. Handling that recognises the horse's ability to identify individuals, remember past interactions and alter future behaviour based on those memories results in improved human-horse relationships and training outcomes, contributing to improved wellbeing.

Respecting equine cognitive abilities and needs allows us to structure their care, training, housing and handling in ways that optimise horse welfare and promote positive psychological wellbeing. Continued research into equine cognition will no doubt further illuminate what environments and interactions are best suited for horses to thrive physically and mentally in an enriched and optimal good life.

Challenges and controversies in equine welfare

The behavioural flexibility of horses leaves them vulnerable to compromised welfare, indeed the horse appears to tolerate restraint in ways that most other species would not. There is a wide range of training, equipment, behavioural and husbandry practices in the equine world that has the potential to compromise horse welfare in the pursuit of human goals. Perhaps by recognising that behavioural indicators of reduced horse welfare are also related to human safety we can create further opportunities to improve both. Changes in horse care and training that result in improved welfare are reflected in positive horse behaviour which in turn offers human safety gains. This challenges the belief that horse riding is inherently dangerous and that riders must accept getting hurt. Not only do practices that result in poor horse welfare harm horses, they also harm riders (Luke *et al.*, 2022).

The use of restrictive equipment and training methods

There is much debate around the use of equipment that has the potential to cause pain, impede normal functions like chewing/swallowing, restrict head movement and prevent horses from expressing discomfort. What we do know is that tight nosebands and riding horses with their heads behind the vertical causes stress and pain. We know that the horse's mouth did not evolve to accommodate a bit and research has shown that since the bit sits on the tongue and the tongue normally fills the oral cavity, there is the danger of the bit being pressed up on to the bony, hard palate and causing discomfort. We know that any equipment that fixes the head position denies the horse's need to move their head forward and backwards, particularly in paces such as the walk and canter. We also know that the position of the head and neck are pivotal in balance, so the effect of head/neck restraint should not be underestimated (Luke *et al.*, 2022).

In essence, welfare can be compromised by the restriction of natural movement as well as by excessive pressure from tack and equipment. Pressure is a function of area so with regard to saddle, bridles, girths and halters, the broader the area that makes contact with the horse the less discomfort it causes; the most everyday consideration in this respect probably being the handler's choice of a rope halter or flat headcollar.

Misinterpretation of behaviour

Several behaviour and welfare researchers have identified that misunderstood horse behaviour is one of the most significant welfare issues facing horses (Horseman *et al.*, 2017). Ridden horse behaviours such as bucking, bolting and spooking are examples of physiologically aroused, reactive behaviour that is regularly misinterpreted by riders as the horse being naughty or disrespectful. This misidentification can lead to harmful escalation in training practices and inappropriate punishment. In reality, reactive behaviour likely signals a mental or physical problem for the horse in terms of being faced with a rider question that they don't understand or cannot cope with performing, whether through physical ability or pain. The development of tools including facial grimace scales (Dalla Costa *et al.*, 2014; Gleerup (2015), the ridden horse pain ethogram (Dyson, 2022) and the equine discomfort ethogram (Torcivia and McDonnell, 2021) highlights a range of behavioural expression now considered to be indicative of pain and therefore, if not addressed, compromised welfare, but were until recently more likely to have been misinterpreted as the horse "refusing" to comply with a request through their own intrinsic recalcitrance.

Demands of some equestrian sports may overwork horses, compromising their musculoskeletal health as well as impacting their hydration/nutrition status and acid-base balance, which reflects the physiological stability necessary for an animal's body to function optimally. The physiological challenges of ridden work highlight the discrepancy between the horse's evolution and what is required of them today by humans. Other than being pushed beyond their physiological limits, potential problem areas include excessive periods without food and water that may compromise circulating blood volume levels, feeding rationed meals that cause fluctuations in gastric acidity, insufficient recovery days after exercise and lack of adequate/inappropriate warming up (McGreevy *et al.*, 2011).

Free-roaming horses spend most of their day with their heads down close to the ground, grazing, moving and subsequently stretching their backs. In contrast, there is evidence to suggest that stabling increases the risk of sore backs perhaps as a result of reduced movement or compromised posture, also the repetitive snatching at hay nets with a high head and neck posture. Whilst being ridden horses are challenged to perform at a higher or more sustained level than their free-roaming counterparts, significantly increasing the movement of the back and predisposing them to orthopaedic injury. Lameness and sore backs are common among ridden horses, but they are largely avoidable. A horse appropriately conditioned and trained to carry riders correctly is likely to avoid many of the injuries that may otherwise be sustained and that includes considering the weight (rider, tack and equipment) the horse is being asked to carry in relation to their ideal body weight. Research has shown that horses asked to carry weights in excess of 20% of their ideal body weight show signs of temporary lameness. In other words, their soundness and therefore their welfare are compromised without due consideration of the loads they are expected to carry. This is why it's so important to understand the forces on the horse's back caused by the addition of the rider. Ridden at walk the average force on the horse's back is at least equal to the rider's weight, at trot it's equivalent to twice the rider's weight and at canter two and a half to three times the rider's weight!

The use of "dominance", "alpha" positions or "leadership", as humans understand and express these terms, are not concepts that relate to a horse's phenomenological understanding of their world. Horses are social animals who mainly interact on a bilateral level (each horse has an individual, and, crucially, fluid, relationship with each other horse) and there is no robust evidence that they have a rank order that encompasses all members of the group. Horse training methods that attempt to dominate and make horses "submit" often precipitate the application and justification of punishment when the horse doesn't comply (ISES, 2017). Apart from the immediately negative effect on the horse's welfare, the wider working relationship is also likely to suffer. The natural and preferred (if available) response of a horse to an aggressive individual is to avoid that individual by moving away. If the horse experiences the rider, trainer or handler as aggressive their main motivation will be to avoid them (Fenner *et al.*, 2019).

Having multiple riders, trainers and handlers can compromise welfare when the use of training methods becomes inconsistent. This inconsistency can lead to confusion, with the horse failing to respond as required or offering inappropriate responses that can result in their behaviour being misinterpreted, harshly corrected or punished. Consider the "problem" horse who has had multiple homes or been passed from trainer to trainer, each with their own approach, methods and system.

Training approaches that have their basis in establishing a cooperative relationship with the horse would appear to be more closely aligned to the social relationships seen in free-roaming herds. It's also interesting to consider that all pressure-based training places the horse in a dilemma. They must respond to some pressures (e.g. from the bit or leg) but habituate to others (e.g. from the girth); we don't want them to habituate to the cues that we give them when we want to control their movement; if they did, these cues (stimuli) would no longer have the effect we require. But fundamentally when being ridden the horse no longer has autonomy over their locomotory responses. If given the choice, would horses prefer not to be ridden since this is accompanied by a loss of control of where and at what pace they are moving? While free-roaming horses have been observed jumping obstacles, especially when moving in canter or gallop, they will instinctively choose to go round obstacles in their path rather than over them. Jumping is an energetically expensive activity and likely, in nature, only used in escape situations rather than for pleasure. However, many of us have experienced the feeling of an apparently obliging, forward-going horse taking off with power, strength and lightness over a fence. What are we to make of that in terms of their emotional experience? Is it negative or positive? How would we know, and what factors would we take into account to draw a conclusion? At this stage in the scientific exploration of horses' affective states there is a wealth of evidence relating to negative experiences; some researchers have recently turned their attention specifically to horses in terms of identifying signs of positive emotional states and experiences, and more of this research is welcomed.

Putting human demands before welfare

What about our own human projections of our goals and desires? We can't leave these out in our considerations of the motivational factors that underlie the activities of this dual species relationship. Equestrian culture frequently prioritises human performance goals, commercial interests or anthropomorphic notions of "respect" over the physical and psychological needs of horses. Knowing/accepting that the horse has choices that might conflict with our own means we must decide what we do if this conflict arises – force, coerce, persuade, stop asking? We are prompted to address our own goals and identity. For instance, we may want to do high level dressage or show jumping, but how do we know how our horse feels about doing these activities? We often hear people talk about their horse loving jumping – how do we know that? What observable behaviour makes us think that, and could there be another interpretation? (Smith *et al.*, 2022). Relevant here is the psychological concept of cognitive dissonance: mental discomfort that results from holding two conflicting beliefs, values or attitudes. Some methods of horse training traditionally focus on delivering results as quickly as possible and without reference to how the horse "feels" about that training. Often very successful trainers have their proven methods and aim to teach students how to do it their way. This can result in those students acting in a way that is not in line with their own beliefs and values (e.g. "I want my horse to enjoy their work and be a partner in our work together, and not just robotically do what they're told"). Another example of cognitive dissonance is the person who has both dogs and horses. They would never dream of using a whip on their dogs (even just tapping) but wouldn't think twice about using a whip to achieve a performance goal with their horse. Similarly with nosebands designed to "close" the horse's mouth; who would strap their dogs mouth shut, knowing it would cause pain and discomfort? The belief exists that strapping one animal's mouth closed is wrong, but that same belief does not come into conscious awareness when applied to the horse and when there is a goal of controlling the horse in order to better the performance. Additionally "everyone does it". Cognitive dissonance, when our actions are not congruent with our values and beliefs, often shows up when we are challenged or questioned about why something like this is ok for horses but not for dogs. We become uncomfortable and defensive whilst trying to justify our actions. However, awareness of our own cognitive dissonance can also prompt us to take positive action to align our beliefs and values with our actions once we can recognise and accept the source of our discomfort.

The more we know about the true nature of horses, who they are and what constitutes a "good life" for them, the more we see examples of where our goals may clash with their opportunity for living that good life. Track systems manage weight for physical welfare but if too restrictive they can challenge mental wellbeing: quite a dilemma when you consider they may score well in the health domain of the five domains model, but less so in the behavioural interactions domain. Riding out alone is more stressful than we typically give it credit for, as is moving horses from yard to yard – even when we do it to provide better living conditions for our horses (or perhaps better facilities for us). At the same time, remember that appropriate habituation (not flooding!), and the horse's great propensity to habituate, can be our friend in this respect.

References

Barnard, C., & Hurst, J. (1996). Welfare by design: The natural selection of welfare criteria. *Animal Welfare*, *5*.

Boissy, A., Manteuffel, G., Jensen, M. B., Moe, R. O., Spruijt, B., Keeling, L. J., Winkler, C., Forkman, B., Dimitrov, I., Langbein, J., Bakken, M., Veissier, I., & Aubert, A. (2007). Assessment of positive emotions in animals to improve their welfare. *Physiology & Behavior*, *92*(3), 375–397.

Dalla Costa, E., Minero, M., Lebelt, D., Stucke, D., Canali, E., & Leach, M. C. (2014). Development of the Horse Grimace Scale (HGS) as a pain assessment tool in horses undergoing routine castration. *PLoS One*, *9*(3), e92281.

Dyson, S. (2022). The ridden horse pain ethogram. *Equine Veterinary Education*, *34*(7), 372–380.

Fenner, K., Mclean, A. N., & McGreevy, P. D. (2019). Cutting to the chase: How round-pen, lunging, and high-speed liberty work may compromise horse welfare. *Journal of Veterinary Behavior*, *29*, 88–94.

Fernandez, E. J. (2024). The least inhibitive, functionally effective model: A new framework for ethical animal training practices. *Journal of Veterinary Behavior*, *71*, 63–68.

Fraser, A. F. (1997). *The Behaviour of the Horse*. Oxford: CAB International.

Gilbert, M., & Gillett, J. (2011). Equine athletes and interspecies sport. *International Review for the Sociology of Sport*, *47*(5), 632–643.

Gleerup, K. B. (2015). An equine pain face. *Vet Anaesth Analg*, *42*(1).

Gov.UK. (2023). *Looking after Horses*. Available at www.gov.uk/keeping-horses

Hemmings, A. J., McBride, S. D., & Smith, N. C. (2004). The putative reward function of equine stereotypic behaviour. *BSAP Occasional Publication*, *32*, 67–78.

Horseman, S. V., Buller, H., Mullan, S., Knowles, T. G., Barr, A. R., & Whay, H. R. (2017). Equine welfare in England and Wales: Exploration of stakeholders' understanding. *Journal of Applied Animal Welfare Science*, *20*(1), 9–23.

Hothersall, B., & Casey, R. (2012). Undesired behaviour in horses: A review of their development, prevention, management and association with welfare. *Equine Veterinary Education*, *24*(9), 479–485.

ISES. (2017). *Position Statement on the Use/Misuse of Leadership and Dominance Concepts in Horse Training*. Available at www.equitationscience.com/equitation

Kelly, K. J., McDuffee, L. A., & Mears, K. (2021). The effect of human–horse interactions on equine behaviour, physiology, and welfare: A scoping review. *Animals*, *11*(10), 2782.

Kiley-Worthington, M. K. (1997). *Equine Welfare*. London: J.A. Allen.

Kiley-Worthington, M. K. (2004). *Equine Education*. Stowmarket: Whittet Books Ltd.

Lesimple, C. (2020). Indicators of horse welfare: State-of-the-art. *Animals*, *10*.

Littlewood, K. E. (2023). The agency domain and behavioural interactions: Assessing positive animal welfare using the five domains model. *Frontiers in Veterinary Science*, *10*.

Long, M., Dürnberger, C., Jenner, F., Kelemen, Z., Auer, U., & Grimm, H. (2022). Quality of life within horse welfare assessment tools: Informing decisions for chronically ill and geriatric horses. *Animals*, *12*(14), 1822.

Luke, K. L., McAdie, T., Smith, B. P., & Warren-Smith, A. K. (2022). New insights into ridden horse behaviour, horse welfare and horse-related safety. *Applied Animal Behaviour Science*, *246*, 105539.

McBride, S. D., & Cuddeford, D. (2001). The putative welfare-reducing effects of preventing equine stereotypic behaviour. *Animal Welfare*, *10*.

McBride, S. D., & Hemmings, A. (2009). A neurologic perspective of equine stereotypy. *Equine Veterinary Science*, *29*.

McGreevy, P. (2007). Stereotypic behaviour in the stabled horse: Causes, effects and prevention without compromising horse welfare. In N. Waran (Ed.), *The Welfare of Horses* (pp. 99–124). Dordrecht: Springer.

McGreevy, P., Berger, J., De Brauwere, N., Doherty, O., Harrison, A., Fiedler, J., Jones, C., McDonnell, S., McLean, A., Nakonechny, L., Nicol, C., Preshaw, L., Thomson, P., Tzioumis, V., Webster, J., Wolfensohn, S., Yeates, J., & Jones, B. (2018). Using the five domains model to assess the adverse impacts of husbandry, veterinary, and equitation interventions on horse welfare. *Animals*, *8*(3), 41.

McGreevy, P., McLean, A., Buckley, P., McConaghy, F., & McLean, C. (2011). How riding may affect welfare: What the equine veterinarian needs to know. *Equine Veterinary Education*, *23*(10), 531–539.

McMillan, F. D. (Ed.). (2019). *Mental Health and Well-Being in Animals*. CABI.

Mejdell, C. M., Jørgensen, G. H., Buvik, T., Torp, T., & Bøe, K. E. (2019). The effect of weather conditions on the preference in horses for wearing blankets. *Applied Animal Behaviour Science*, *212*, 52–57.

Mellor, D. J. (2016). Moving beyond the "five freedoms" by updating the "five provisions" and introducing aligned "animal welfare aims". *Animals*, *6*(10), 59.

Mellor, D. J. (2017). Operational details of the five domains model and its key applications to the assessment and management of animal welfare. *Animals*, *7*(8), 60.

Mellor, D. J., & Beausoleil, N. J. (2015). Extending the 'Five Domains' model for animal welfare assessment to incorporate positive welfare states. *Animal Welfare*, *24*(3), 241–253.

Mellor, D. J., Beausoleil, N. J., Littlewood, K. E., McLean, A. N., McGreevy, P. D., Jones, B., & Wilkins, C. (2020). The 2020 five domains model: Including human–animal interactions in assessments of animal welfare. *Animals*, *10*(10), 1870.

Mellor, D. J., & Reid, C. S. W. (1994). Concepts of animal well-being and predicting the impact of procedures on experimental animals. In Baker, R. M., Jenkin, G. & Mellor, D. J. (Eds.), *Improving the Well-being of Animals in the Research Environment* (pp. 3–18). Glen Osmond: Australian and New Zealand Council for the Care of Animals in Research and Teaching.

Nordenfelt, L. (2006). Biological theories of animal welfare. In *Animal and Human Health and Welfare: A Comparative Philosophical Analysis* (pp. 54–70). Wallingford: CABI.

Rochais, C., Henry, S., Fureix, C., & Hausberger, M. (2016). Investigating attentional processes in depressive-like domestic horses (Equus caballus). *Behavioural Processes*, *124*, 93–96.

Smith, R., Furtado, T., Brigden, C., Pinchbeck, G., & Perkins, E. (2022). A qualitative exploration of UK leisure horse owners' perceptions of equine wellbeing. *Animals*, *12*(21), 2937.

Torcivia, C., & McDonnell, S. (2021). Equine discomfort ethogram. *Animals*, *11*(2), 580.

Waran, N., McGreevy, P., & Casey, R. A. (2007). Training methods and horse welfare. In *The Welfare of Horses* (pp. 151–180). Springer.

Wemelsfelder, F. (2008). Qualitative Behaviour Assessment (QBA): A novel method for assessing animal experience. *Proceedings of the British Society of Animal Science*, *2008*.

Wemelsfelder, F., Hunter, T. E., Mendl, M. T., & Lawrence, A. B. (2001). Assessing the 'whole animal': A free choice profiling approach. *Animal Behaviour*, *62*(2), 209–220.

Wilkins, C. (2023). *Applying the Five Domains Model to the Welfare Assessment of Sport and Recreation Horses*. Available at www.openlearning.com/une/courses/une-equine-course/?cl=1

Part 4

Behaviour consulting

Chapter 8

Diagnosing behaviour problems – how to assess a behaviour problem and identify its likely causes

Legal and professional aspects of behaviour consulting

The first part of the framework that behaviour consultants are bound by is the Veterinary Surgeons Act (1966), which limits the activities that may be carried out by those who are not registered vets (RCVS, 2023). This includes diagnosis of diseases and injuries, giving advice based upon that diagnosis and the medical treatment of animals. There are a number of exceptions (outlined in Schedule 3 of The Act) which allow minor medical treatment to be carried out by:

- the animal's owner, member of their household or their employee
- the owner or person engaged in caring for animals used in agriculture
- registered vet nurses, student veterinary nurses and veterinary students
- registered farriers
- anyone administering emergency first aid to save life or relieve pain or suffering

As paraprofessionals (a person trained to assist a professional person, in this case the vet), equine behaviourists in the UK who are registered with the main professional bodies for clinical animal behaviour in the UK are required to work under the referral of the horse's vet, meaning that the vet gives consent for the practitioner to consult in respect of the behaviour problems reported by the horse's owner (FAB Clinicians, 2023). In effect, the horse is placed under the duty of care of the behaviourist for the purposes of treating them behaviourally, until the behaviourist discharges the horse back to the vet. During this time the practitioner and the vet will stay in regular communication to ensure that all interactions between health and behaviour are addressed during the consultation period.

The role of a clinical animal behaviourist (CAB) is to look at all the possible causes of the unwanted behaviour reported by the client and through their knowledge and fact-finding identify and treat the actual cause (CCAB, 2023; ABTC, 2021). In essence they make a behavioural "diagnosis" and recommend a treatment/behaviour modification plan to rectify it. You may notice that some clinicians prefer to use terms like assessment or case formulation rather than diagnosis to avoid any potential conflict with the Veterinary Surgeons Act, and in reality, within a biopsychosocial framework, "formulation" is a suitable term, taking into account the multifactorial causes involved (Campbell and Rohrbaugh, 2013). Clinicians explore the behaviour of the horse by taking a detailed "whole of life" case history combined with direct observation

DOI: 10.1201/9781003581390-12

in order to provide a broad analysis of the horse's mental wellbeing, which may include trauma and stress-related issues for the individual animal in their history or their current situation/environment. In order to address any unwanted, inappropriate, problematic or dangerous behaviours a plan is put in place to modify the behaviour (Carroll *et al.*, 2023).

It is a membership requirement of these professional bodies that all consultants work as part of a vet-led team, not least because of the link between some medical conditions and behavioural indicators, and agree to a code of conduct which generally includes some form of the following:

- to only take on cases with written referral from the horse's registered vet
- to ensure any actual or potential medical causes for problem behaviour are identified/treated by a vet prior to behavioural treatment
- to ensure that diagnosis and treatment of medical disorders that may be associated with a problem behaviour is carried out by a vet
- to discuss any medical treatment of a behaviour case with the referring vet by referencing the current and established scientific literature
- do not in any way advise the client or the vet of specific medical or psychopharmacological treatment
- do not give any specific form of physical health-related advice without consulting the referring vet
- recognise the limits of their competence, do not attempt to practise beyond them and refer on where appropriate

The CAB may suggest the use of medication to the referring vet as part of case discussions, which is in line with the Veterinary Surgeons Act, but the decision about which, if any, medications to use and the prescribing of those medications remains the responsibility of the referring vet. In practice, this means that if the CAB suspects the behaviour may have an underlying medical cause they are likely to use phrases in their report such as:

- the behaviour this horse is exhibiting has been found to be consistent with physical pain or discomfort . . . (followed by references from the literature that support this)
- there is a paper by . . . that suggests this type of behaviour can be indicative of this physical condition, I'd be grateful for your thoughts
- it would be helpful to eliminate pain, disease or dysfunction as differentials for . . . behaviour and I would appreciate your advice on how best to do this
- you may wish to conduct blood, faecal or other tests for any digestive, hormonal or other abnormality

The importance of this approach cannot be overstated from a professional indemnity point of view. Whilst it is a requirement for all clinical animal behaviour consultants to have both professional indemnity insurance and public liability insurance, this will not protect the CAB if they involve themselves in aspects of diagnosis, advice or treatment which are outside their professional remit.

The animal welfare legislation described in chapter 7 is also relevant here, and health and safety at work (HASAWA) regulations establish a good practice framework for practitioners to conduct risk assessments in respect of their cases.

Working under vet referral

In the early stages of arranging a consultation the CAB will ask for details of the vet that the horse is registered with. Good practice is for the behaviour consultant to email the vet seeking consent for a referral, outlining the nature of the unwanted behaviour reported by the owner and requesting the horse's medical history. The main professional behaviour bodies have approved referral forms for animal behaviour cases which can be downloaded. The medical history is particularly important and helps the CAB to understand the horse's full behavioural history. Any periods of pain, illness or hospitalisation at any point in the horse's life can influence aspects of their behaviour. Once consent has been given, the referring vet should be kept informed of the progress of the consultation including an initial written report detailing what was found and the proposed treatment plan. This may then be followed up with updates on any developments as appropriate and eventually the case outcome. It might be helpful to think of this process in the same terms as a hospital consultant; just as a GP would refer their patient to the specialist consultant, the vet refers the horse to a behaviour consultant for investigation and treatment. Once the CAB has finished their consultation the horse is discharged back into the full care of the vet.

Example vet referral email

Dear Sir or Madam,

I trust this email finds you well.

I should be grateful for your kind referral to your client [client name], owner of [horse name], [breed], d.o.b. dd.mm.yyyy

[Owner name] has requested my services in respect of [horse name]'s behaviour of:

[use the wording the owner has used to describe the behaviour, on their questionnaire – (what is the behaviour problem you are seeking help for?)]

I seek your kind referral to consult in respect of the reported behaviour problem(s), in accordance with the requirements of the Animal Behaviour and Training Council.

Please could you complete the attached referral form and return it to me at your convenience?

If you could also send a copy of [horse name]'s medical history that would be very helpful.

Many thanks for your kind attention.

Elements of a case history

The first step to developing a case history is fact finding. Alongside the medical history from the vet, ask the owner to complete a questionnaire about the things they have observed in the horse, how they are currently managed and trained, any previous history they have available and what the current behavioural problem is that they're seeking help for (Table 8.1). Questions might include:

- the nature of the behavioural problem
- general management currently
- details of current riding and handling
- behaviour with other people
- the horse's background
- any history of aggression to horses/humans
- yard/field environment
- turnout details and turnout group
- general health and wellbeing
- observed behaviours – by which you will be able to identify natural ethogram behaviours and any abnormal behaviours within a specific context

Armed with this information the CAB then undertakes a visit to observe the horse and owner interacting, ask further questions and clarify any details from the questionnaire and medical history. Ideally the CAB observes the horse and owner doing all their normal routine activities even if they are not part of the problem behaviour as they can often provide a helpful insight into the relationship. Some owners may want to demonstrate the problem behaviour on this visit; however, it is neither helpful nor ethical to put the horse into a position of compromised welfare or a fearful state just to demonstrate the unwanted behaviour. In all likelihood the CAB will be familiar with the types of behaviour being described and won't need to see it manifested in this particular horse. At the end of the visit the CAB will have assessed a number of factors that may be contributing to the behavioural problem.

Assessment of environmental and management factors

The CAB assesses how well the horse's needs are being met in terms of meeting the horse's basic species-specific needs; the five welfare domains (nutritional conditions, physical environmental conditions, health conditions, behavioural interactions, mental state) can be a helpful starting point or framework for this. The main purpose is to identify where positive and negative circumstances may impact on the horse's affective state and highlight any areas of imbalance or disruption.

Assessment of relationships

In terms of relationships the CAB is looking to understand how the horse behaves with the different people they interact with. For instance, the farrier, vet, yard staff, trainer and physio or other body workers as well as the owner themselves. Studies have shown that people are actually quite poor at recognising fear and pain in horses, particularly the more subtle signs (tension in face and avoidance behaviours such as hesitant gait) and these are often misinterpreted or mislabelled (Bell *et al.*, 2019). This may explain why a behaviour specialist is called in only once more overt (and often dangerous) signs such as rearing, bolting or spinning start to occur.

Table 8.1 Areas of Questioning in the Consultation

Turnout	Stabling	Feeding
How much turnout does the horse have – size/time? On their own/pair/group? Do they have a preferred field mate? How long have they been companions for? What is the turnout/field like? Has anything changed in the last two years?	How often is the horse stabled? What size is the stable? Where is the stable located? What is the stable made of? What bedding is used and how is it mucked out? What can the horse see from the stable? Who are they stabled next to? How is water given? Has anything changed in the last two years?	What is being fed and when? How is food given – bucket/feeder/floor? Hay/haylage – how much and how often? How is forage given – net/floor? Any supplements being used? How well does the horse eat? Is food being restricted or monitored? Has anything changed in the last two years?

At this point, and also quite early on in a CAB's interaction with an owner, it is worth talking about a horse's behaviour threshold. In a behaviour threshold the horse crosses from one emotional state to another. Perhaps from calm to excited or from relaxed to fearful. Consider the horse who worries about the scary corner of the arena. They move from appearing calm and relaxed to feeling tense, slowing down to avoid going towards that area, they stop listening to the leg aid, spin or become difficult to ride. When this happens it is a clear indication that the horse has gone over threshold. Often owners think of over-threshold horse behaviour as rearing, napping or bolting, but there are other more subtle ways that a horse tells us they are over threshold or moving from a comfortable emotional state into an uncomfortable one:

- shutting down or appearing to freeze
- being overexcited
- being so distracted that they are unable to pay attention to what is being asked
- lip licking and chewing
- yawning
- nipping
- shaking their head
- rubbing their head on something or rubbing their nose on their knee
- stop taking food treats
- stop interacting
- tension around the mouth
- looking away
- stiffness/slower movement
- raising tail away from rump
- defecating

From an owner's perspective when their horse is over threshold it generally means that they are behaving in a way that the owners don't like. From the horse's perspective it can often mean that they are experiencing a state of anxiety, fear or distress. By helping the owner understand the concepts of threshold and the things they can do to support their horse to stay under threshold the better the chance of a successful outcome to the treatment plan (Rogers and Bell, 2022).

You can find a client handout on "Your horse's behaviour threshold" at Appendix 2.

Assessment of training and handling factors

Once an understanding of behaviour thresholds has been established the consultant (and owner) can look at how the horse is trained and handled in order to gain insight into how the horse might feel about these interactions and the impact they are having (Zeitler-Feicht, 2004). Does the way the horse is handled, trained or interacted with put them over threshold?

Another useful indicator about how the horse is feeling, in other words their emotional state, is how the horse behaves when being caught, led, groomed and tacked up. Horses evolved to do most of their communication via body language. The movements of different parts of the body can be very subtle and even just a shift of weight, a slight turn of the head or tensing up conveys information. The different expressions of body language can be broken down into three categories – approach, avoid or neutral (Table 8.2). Approach behaviours are those that decrease distance from something (a person, another horse, object or situation). They communicate that the horse is feeling friendly/confident/interested/curious. Avoidance behaviours aim to increase the distance from something. They communicate that the horse is feeling anxious/fearful and signal "I know what happens next and I'm worried about it". Neutral behaviours indicate relaxation and are most of the behaviours described in the equine ethogram in chapter 2.

Table 8.2 Approach, Avoidance and Neutral Behaviours

	Approach	Avoid	Neutral
Being caught	Walks towards person Soft face, ears pricked	Walks away Turns head away Turns bum towards person	Continues grazing/resting Hanging out/loafing
Being led	Walks with person Interacts with person leading Soft face, ears pricked	Slowing down Planting Nudging or nipping	Soft musculature Loose hanging tail
Being groomed	Sings of pleasure if you find an itchy spot	Turns head away Nibbling persons back while feet being picked out Fidgeting	Standing calmly Soft face Loose hanging tail Soft musculature
Being tacked up	Looking towards person bringing tack Soft face Interacts with person tacking up	Head disappears in stable as tack comes out Standing in furthest corner of stable Throwing head up Fidgeting Moving away from saddle Kicking or biting as girth done up	Standing calmly Soft face Loose hanging tail Soft musculature

It's important to note that when avoidance behaviour is observed while being tacked up, then as well as fear or anxiety, pain should be considered, either experienced at the time or because being tacked up is a reliable predictor of "what happens next", likely being ridden and potentially experiencing pain or discomfort. A 2022 study by Kjærulff and Lindegaard (2022) found that changes in horses' ridden performance or behaviour are more often caused by various pain-related conditions than attributed to "true" behavioural problems. Tacking up could also predict something emotionally difficult for the horse such as leaving companions to go for a hack, or being ridden somewhere they are scared of, or a schooling session where they have learned will experience confusion/frustration.

Differential diagnosis

The behaviour clinician now needs to conduct a differential diagnosis. This involves looking at all the possible causes of the behaviour problem and using the data gathered to strip away the least likely explanations until arriving at a behavioural "diagnosis", assessment or formulation. In doing this the CAB identifies the contributory predisposing, initiating and maintaining factors, keeping in mind that there may be a medical element in any or all of these factors. The first step is always to rule out general medical conditions that can cause behavioural signs. Clinical behaviourists are not experts in assessing horses for medical conditions and cannot legally or by their code of practice operate outside their area of competence. This is why they work in liaison with the referring vet who can establish any medical disease or dysfunction before the CAB assesses for behavioural disorders.

Predisposing factors

Unwanted behaviours are often "threat avoiding reactions" that are influenced by certain predisposing factors. The CAB must ask, what was it that predisposed this horse to this problem? Could there be a genetic or breed related element? There are some more sensitive breeds that tend towards panic and over-reaction whilst others tend towards planting and unresponsiveness. Some horses may react with panic if they are given no possibility of escape. They may also react similarly when asked to perform tasks that they don't understand. Equally, horses with less than ideal conformation are often unable to perform the tasks/movements that are being asked of them. Some management practices in particular predispose to unwanted behaviour: high concentrate feeds, lack of exercise, lack of social contact and a barren environment can provoke problems such as rearing, bucking and bolting. Are there any underlying medical conditions contributing to the behaviour as a predisposing factor by causing pain/discomfort, or is problematic behaviour contributing to a medical condition?

Initiating factors

What was it that actually initiated the problem in this horse? What was the first thing that happened that started the behaviour presenting? It could be a single event or a combination of event plus environment, or a significant change in the horse's life that triggered the behaviour.

Maintaining factors

What is it that is maintaining the behaviour problem now? Why is it still happening? It could be a maintaining factor such as learned behaviour (continuing to be reinforced), or the conditions that gave rise to the problem initially are still present (Figure 8.1). Perhaps the owner is inadvertently doing something that is perpetuating the behaviour. A lack of understanding of how horses learn can result in the unconscious training of unwanted behaviours. For instance, a horse pulls their foot away from the farrier when they are being shod. The owner, with the intention of distracting or calming the horse, gives them a handful of chaff. Unfortunately from the horse's perspective the "pulling their foot away" behaviour is rewarded and consequently the horse is likely to repeat the behaviour in future. It's the same process with a horse kicking their door, rubbing their rump against the stable wall, or pawing. For a bored horse the resulting reprimanding attention, even a shout of "oi!", can be reinforcing and thus maintains that behaviour.

In addition, the CAB needs to consider context, emotional and motivational aspects relating to the unwanted behaviour. Where and when is the behaviour happening? What is the underlying emotion (e.g. fear/anxiety)? And what is the horse's motivation? Are they trying to increase distance between themselves and something they are concerned about? The answers to these questions provide valuable insight into why the horse is exhibiting the behaviour. Fear and pain are the major causes of unwanted behaviour. Aside from medical issues, pain can be caused by improper riding/training techniques, ill-fitting tack/equipment, harsh bits, harsh hands and the inappropriate use of training aids/equipment. Fear is an innate reaction of a flight animal to unknown situations and is enhanced by conflict situations, deprivation, bad experiences and pain. Typical reactions shown by fearful horses include herd boundness, bolting and shying. If

Figure 8.1 Continuing to attempt to administer wormer to this horse is likely to be a maintaining factor in their resistant behaviour. Photograph by Chelle129.

they are restrained by the owner/handler with force, this can escalate in the form of bucking or rearing.

The key here is to consider multiple explanations in a systematic way, collaborating with the vet to rule out medical issues and gathering a holistic understanding of the roots of the behaviour in order to develop an effective, personalised behaviour treatment plan. The treatment plan is specifically tailored to the individual horse and the goals of the owner in terms of what they want to be able to do, and for the consultant there may be some expectation management communication involved here. The treatment plan includes explanations of the likely causes and factors maintaining the unwanted behaviour. Treatment recommendations may include changes in handling, management, environment and training modifications to suit the individual horse. Medical treatment may be a consideration, in consultation with the vet.

The principle of parsimony

The principle of parsimony, also known as Occam's razor, suggests that among the theories fitting the data equally well, scientists should choose the simplest explanation. The process of differential diagnosis allows the consultant to identify all potential causes of the behaviour, strip out any causes that are completely out of the question (for example, perhaps because a medical investigation has ruled them out or the saddle has already been checked), and focus on the one or two most likely causes.

Useful resources to leave with owners

At the end of the visit it is often helpful to provide the owner with some useful resources/activities that they can read or implement with their horse while they are waiting for the full report and treatment plan from the CAB. It's important not to overwhelm the client with heaps of information, but a few things they can be doing in the interim is a great starting point in preparation for implementing the full treatment plan. You can find a selection of useful resources in the Appendix:

- "Being a bringer of good things" Appendix 3
- "How to enrich your horse's environment" Appendix 4
- "Target game" handout Appendix 5
- "Teaching 'Find it!' to your horse" Appendix 6

Example case reports and an example client report can also be found at Appendices 7, 8, 9 and 10, and an example shaping plan at Appendix 11.

Visual illustrations of what a calm, relaxed horse looks like in contrast to indicators of stress or anxiety are also helpful, as in Figures 8.2 and 8.3.

Figure 8.2 A comparison of facial expressions: Drago, relaxed. Photograph by Maïa Nootens, Equine Photographer.

Figure 8.3 A comparison of faces: stress and tension. Photograph by Anke van Wyk.

References

ABTC. (2021). *ABTC Standard Clinical Animal Behaviourist*. Available at https://abtc.org.uk/wp-content/uploads/2021/12/ABTC-Standard-Clinical-Animal-Behaviourist-November-2021.pdf

Bell, C., Rogers, S., Taylor, J., & Busby, D. (2019). Improving the recognition of equine affective states. *Animals*, *9*(12), 1124.

Campbell, W. H., & Rohrbaugh, R. M. (2013). *The Biopsychosocial Formulation Manual: A Guide for Mental Health Professionals*. Routledge.

Carroll, S. L., Sykes, B. W., & Mills, P. C. (2023). Understanding and treating equine behavioural problems. *The Veterinary Journal*, *296*, 105985.

CCAB. (2023). *What Is Clinical Animal Behaviour?* Available at www.ccab.uk/what-is-clinical-animal-behaviour

FAB Clinicians. (2023). *Code of Conduct*. Available at https://fabclinicians.org/code-of-conduct/

Kjærulff, L. N. R., & Lindegaard, C. (2022). Performance and rideability issues in horses as manifestations of pain: A review of differential diagnosis and diagnostic approach. *Equine Veterinary Education*, *34*(2).

RCVS. (2023). *Treatment of Animals by Unqualified Persons*. Available at www.rcvs.org.uk/setting-standards/advice-and-guidance/code-of-professional-conduct-for-veterinary-surgeons/supporting-guidance/treatment-of-animals-by-unqualified-persons/

Rogers, S., & Bell, C. (2022). Perceptions of fear and anxiety in horses as reported in interviews with equine behaviourists. *Animals*, *12*.

Zeitler-Feicht, M. H. (2004). *Horse Behaviour Explained: Origin, Treatment and Prevention of Problems*. London: Manson Publishing.

Chapter 9

Treatment planning – safety and situation

Safety

The essential prerequisite to behaviour modification, which we will cover in the next chapter, is keeping people and other animals safe and recognising that the client horse also needs to feel safe in both ethological and emotional terms. In this sense it's clear that a non-vet behaviourist is providing a service of behaviour therapy. Behaviour-related issues are common in horses and many "unwanted behaviours" pose safety concerns for human handlers, riders, owners and other animals, as well as welfare concerns for the horse. It may sound like common sense but in these initial stages it's important to be clear with owners about not putting themselves at risk. For instance, if you are being consulted about a horse that bolts, advise the owner not to ride them in situations where bolting is likely to happen or has happened in the past. Equally, if their horse is problematic in the arena, don't school. This is not a time for "riding their horse through it". The bottom line is, don't do the things that the owner knows is likely to result in unsafe behaviour, whether that's when the horse is being either ridden or handled. Not only is it putting the human at risk, it's also strengthening the unwanted behaviour by facilitating its repetition.

At this point it's worth mentioning trauma in the context of safety. In most cases, following a stressful event, the body's physical and psychological processes are disrupted but soon return to a baseline homeostasis – in other words they return to their normal day to day levels of functioning. However, studies in both humans and animals have shown that following severe, prolonged or repetitive psychological stress, that baseline level of functioning is not re-established and the body continues to operate as if the trauma is a persistent and ongoing part of their daily life. We know that trauma produces actual physiological changes in the body including a recalibration of the brain's alarm system, an increase in stress hormone activity and an alteration in the system that distinguishes relevant information from irrelevant information. Hence people who have been traumatised become hypervigilant to threat and have trouble learning from experience. Research has shown that this is true for dogs and in my experience this is clearly observed in horses too. In cases where trauma is suspected, it's helpful to adopt an "all bets are off" type of approach, recognising that the progress you might typically expect to see in terms of desensitisation and/or counterconditioning is not likely to follow a linear trajectory; relapse is common and even different or unexpected behaviours may appear that superficially might not appear related to your assessment or treatment plan.

What is mental or emotional trauma? Any overwhelming event, series of events or set of circumstances that has overwhelmed an individual's ability to process and integrate something

DOI: 10.1201/9781003581390-13

that has happened to them and that has lasting adverse effects on their functioning and wellbeing. Probably the most helpful way to look at mental or emotional trauma is that it is subjective. How an individual perceives an event or situation influences their degree of trauma. Two individuals, horse or human, can experience the same event and one may be traumatised while the other is not. So it's important to distinguish between a potentially traumatic event, often referred in the literature as a PTE, and an individual's response to that potentially traumatic event which may or may not result in mental or emotional trauma. In essence, trauma is not the event, it is the individual's response to that event.

Potential causes of trauma in the horse:

- poor weaning practice
- unpredictability of a new home
- patterns of insecure attachment (multiple homes/handlers/riders)
- inadequate opportunities to socialise/interact with other horses
- prolonged restricted access to forage
- starvation
- restricted ability to move/prolonged confinement or restraint
- conditioned negative associations (with handler/context)
- flooding – (possibly unintentional) by owners/handlers/trainers
- abuse/neglect – whether intentional or not
- loss of a close companion (death/being sold)

Given the way that horses are bought, sold and change hands, it can be difficult to identify those whose circumstances may predispose them to effects of trauma. However, whether trauma is suspected or not, the first step in any treatment plan is always the same – to ensure the horse feels safe and that they are given the time they need to allow their physical and psychological processes to work to return the body and brain to homeostasis. Any "unwanted" behaviour being displayed is very often just the tip of a behavioural and emotional iceberg, and in order to be able to support the horse to change their behaviour we need to focus on what lies beneath – in other words their general wellbeing. As discussed in chapter 7, wellbeing is multifaceted, consisting of physical, mental, social and environmental factors which contribute to the horse's emotional balance. By reducing stress and improving the physical, social and environmental factors identified in the differential diagnosis, owners give their horses the best chance to become more relaxed, happier and resilient.

Behavioural first aid

All horses need to feel safe and secure as an essential underpinning of all the other elements in their life (think Maslow's hierarchy of needs in humans). Ideally this needs to be strongly established before moving on to a behaviour modification programme. To this end the initial focus of any treatment plan is behavioural first aid, helping to reset the horse emotionally by building a strong basis of safety and predictability, ensuring nothing unpleasant happens to them and avoiding the stress triggers that caused the unwanted behaviour in the first place. In practice this means avoiding situations where the problem behaviour occurs, including forcing the horse to do things they are clearly not happy about; rethinking field companions if aggression is a problem; stopping the use of punishment (shouting, smacking, lead rope jerking, backing up, kicking); providing an opportunity for the horse to decompress (a holiday ideally turned away in a field with reduced human contact) and meeting their ethological needs. This starts the process of decoupling the connections in the horse's brain where they learned to feel anxious or fearful, causing them to react aversively to objects or experiences they found worrying or unpleasant. An "only good things" approach starts to build the feeling that they are safe, can trust their owner and therefore remain calm.

This is the phase that may meet with the most owner resistance. People like to be "doing" something to fix the problem. Gone, for the most part, are the days when horses were turned away for a prolonged period of time perhaps because it was the off season of the sport or because of the dark winter nights when most didn't have access to floodlit arenas, or indeed because they were presenting with some kind of behavioural difficulty. The modern-day horse has very little opportunity to decompress and many owners (and often practitioners too) prefer to jump straight into cooperative care/desensitisation/training without providing the horse with any sort of mental break to reset their baseline emotional balance. Many owners feel they are giving their horses a "stress holiday" if they've stopped riding but are still bringing them into the stable every day for grooming and a bit of liberty work in the arena. The longer you can give the horse away from human-involved activities at this stage the better.

The next step on the treatment plan is relaxation. The aim here is to develop the horse's ability to relax more often and in a wider range of situations by adding positive horse-centric activities to their daily life. This stage incorporates some easy, pleasant exercises and enrichment that will develop stronger coping abilities in different situations and will also rebuild resilience so that they are no longer motivated to perform unwanted behaviours. In my practice I typically include activities such as Rachaël Draaisma's scentwork (Draaisma, 2020), environmental enrichment, target games for horses and teaching the Find It! game. The key to all of these activities is to ensure that it is a pleasant experience for the horse and they remain below threshold. It is not a quick process. This kind of behavioural and emotional change needs to take place at a cellular level in the horse's brain and it takes time for that to happen. Existing neural connections that underpin the motivation for the unwanted behaviour need to weaken and new connections need to develop that support the new and different emotional state of calm relaxation. A key mood-benefitting brain molecule called brain derived neurotropic factor (BDNF) is one of the reasons why enrichment and opportunities to engage in species-specific behaviours are so important in a horse's daily life. Expression of the BDNF gene has been found to be reduced by chronic stress, inflammation and ageing, and conversely it's been found to increase with physical and mental activity including aerobic exercise, positive social interactions, mental stimulation and environmental enrichment.

It's also important to manage expectations. The resetting and relaxation process is a bit like a game of snakes and ladders. When your client is making progress they're going up the ladder to success. But sometimes something happens that causes the unwanted behaviour to reappear and then they're sliding back down the snake! This is only a temporary setback; providing they stick to your treatment plan they will be back climbing the ladders! The reality is that progress is not likely to go upwards in one consistent direction; it's going to be up one day and down the next and it may even change from hour to hour in particularly complex trauma cases.

Approaches to response management

At this stage in the treatment plan we are aiming to prevent the unwanted behavioural response in the horse, but it can be tricky to think about how *not* to do something. For instance, if the horse is struggling with being shod and the farrier visits every four weeks, how do we maintain foot health, manage the horse's response to this situation and ensure that unwanted behaviour isn't being reinforced? The first question might be, does the horse need shoeing or a farrier visit every four weeks? Or, if you give the horse a haynet, bucket feed or access to a lick treat, does that change their behavioural response to having their feet worked on (in that moment, it's important to recognise)? In extreme cases, would a small amount of medication break the pattern of behavioural responses and make it safer for those involved, both horse and human? This last consideration is of course one to refer back to the vet and is an example of the importance of working as part of a vet-led team in clinical behaviour practice.

If the horse is behaving aggressively, or is being bullied in the field, how do you prevent that happening whilst still meeting the horse's ethological needs for socialisation? Often on yards this situation results in the horse being turned out on their own; however, with a bit of thought perhaps a suitable companion could be found, or a reorganisation or introduction of turnout companions undertaken to prevent the aggression or remove the bullying whilst still ensuring the horse has companionship. If the horse cannot hack out on their own safely, then hacking out with a trusted companion might prevent a napping/rearing/spinning response. At this stage we are not looking to modify/retrain the behaviour (and with it, the underlying emotion), simply to prevent the response happening in the first place. This stage involves being creative with our approach so the horse is not put in a situation that triggers the unwanted behaviour and without restricting their ability to perform the natural behaviours on their ethogram. Our aim is to expand their daily experiences in positive ways but still prevent the unwanted behavioural responses by addressing any contributing management/environmental factors identified in the differential diagnosis.

Promoting adaptive species-specific behaviours

Changing management and environmental factors are key to promoting adaptive species-specific behaviours. What this means in practice is highlighting the ethogram behaviours (see chapter 2 – equine ethology) that the horse currently is unable to do, that it would be helpful from a wellbeing point of view for them to be able to engage in. This might include more turnout, or turnout with conspecifics (Figure 9.1).

Figure 9.1 Turnout with familiar friends meets one of the essential species-specific needs. Photograph by Rita Kochmarjova.

It might mean enriching their environment, whether in the stable or in the paddock. Perhaps encouraging more movement, more foraging or the opportunity to browse and be selective about what they choose to eat, even to the extent of providing some choice about whether to drink from an automatic waterer or a bucket, or from a blue bucket rather than a green one. A 2020 study investigated preference for different coloured buckets and found that significantly more water was consumed from turquoise coloured buckets than other colours. In addition, it was observed that light-toned buckets (turquoise, light blue) were preferred to dark ones (green, yellow, red).

It's important to reiterate that promoting species-specific behaviours needs to be tailored to the individual horse. It's all very well saying that all horses should be turned out 24/7 or that all horses should live in large herds but that won't always be the right thing for that individual horse, at that particular time, in that particular environment. As an example, an owner thinking they are doing the best thing for their horse may move livery yards so they can go from three hours turnout a day to living outside 24/7 in a group. That horse could actually end up more stressed, depressed, even more susceptible to injury. Why? Perhaps the owner didn't take into account that on the first yard with restricted turnout their horse was stabled next to their friend, they always went out in the field together, hacked out together, went to the same competitions together and had never really been apart for several years. Suddenly one of these firm companions is taken away from what and who they know, where they feel safe and secure from their long-time friend, in order to be turned out with a herd of new and unfamiliar horses. Sometimes it is appropriate to make big changes, but sometimes it's enough to make small adjustments that allow for more opportunity to promote appropriate behaviour within the current setup.

Figure 9.2 Taking a horse or pony on an exploratory "nature walk" is a recommended enrichment activity. Photograph by Robert Petrovic.

Figure 9.3 Toy box enrichment activity. Photograph by Joanne Hogg.

In essence the treatment planning process thus far has done the following:

1. Undertaken a differential diagnosis to identify all the things that could be contributing to the unwanted behaviour. Perhaps the horse isn't feeling safe or hasn't got the opportunities to express species-specific behaviours or is experiencing chronic stress and anxiety in some way. Perhaps there's a disconnect in the relationship between the horse and owner or perhaps there's an underlying medical condition.
2. Hit the reset button. Identified the immediate things that we are going to change. This might include training, management or environmental factors as well as stopping doing things that are causing/triggering the unwanted behaviour.
3. Built the buffer. Horses operating at the limit of their behaviour threshold are on the edge of coping. By hitting the reset button the horse is given time to return to their baseline homeostasis. This provides a buffer between the normal functioning of the horse and their behaviour threshold. In this space we can put in enrichment activities, such as those shown in Figures 9.2 and 9.3, before we prepare to reintroduce the horse to the aspect of their life that was causing them a problem.
4. Rebuilding. Once the behaviour consultant and the owner feel that the horse is in a better place emotionally, showing a regular and frequent relaxed outlook, we can move onto the next step, in the next chapter – behaviour modification.

Reference

Draaisma, R. (2020). *Scentwork for Horses*. London: CRC Press.

Chapter 10

Treatment planning – behaviour modification

Modifying behaviour

A typical overall objective in any behaviour modification programme (BMP) is to change how the horse feels about the situation that is currently causing them difficulty, so that they can engage with human-directed routines and activities with calm and relaxed engagement as in Figures 10.2 and 10.3 (McGreevy, 2012; Mills and Nankervis, 2013). For some horses, it is clear that this is not possible and that there is a mismatch between caregiver goals and what the horse is capable of adapting to, whether physically or mentally. Through the behaviour therapy process we aim to move the horse from a negative emotional state to a positive emotional state, in the situation that currently challenges them, so that it becomes something that the horse experiences as pleasant rather than aversive. It's important that the caregiver adopts a mindset of safely avoiding and ethically preventing unwanted behaviour and reinforcing the behaviour they do want. If they need to prevent the horse from doing something, this must be done in a way that avoids anxiety, frustration or confusion. The underlying principle is to prevent unwanted behaviour humanely without causing the horse any stress and to reinforce and condition the behaviours that form part of the BMP (Zeitler-Feicht, 2004; Hothersall and Casey, 2012).

When working with clients the BMP needs to relate back to the goals they identified at the start of the consultation process and be continuously revisited throughout the training process by having regular follow-up progress review sessions. In this way the BMP is broken down into smaller action plans which the client goes away to work on for an agreed period of time. The progress is then reviewed before moving onto the next action plan in the BMP or in some cases going back a few steps if the horse is struggling. Sometimes the client's goals or their outlook change as a result of working through the BMP, which is another reason it is so important to keep checking in to ensure you are both still working towards the same goal, even if that changes during the time you and your client are working together. In some cases a client may have contacted a behaviour consultant because they've experienced problems that they've been unable to resolve and have made it worse. In this scenario, once the horse is back to being able to meet their owner's expectations in a more balanced way and coping better with life in general, the client and the CAB may agree an end point to their work together, after which the client goes back to their regular trainer.

DOI: 10.1201/9781003581390-14

Desensitisation and counterconditioning

Fears, anxiety, phobias and aggressive displays are all features of behaviour problems that involve arousal, negative emotional responses and feelings (Rogers, 2018). When the problem is caused by how a horse feels about an object, person or a situation, it isn't enough to just teach them an alternative behaviour; instead, it's more effective to change the way the horse feels about whatever it is that is putting them over their arousal threshold. This emotional change has a positive effect on the underlying reason for the behaviour problem. Systematic desensitisation and counterconditioning (DS/CC) are often used together for the best effect and are considered the equivalent of gradual exposure therapy in humans. Gradual exposure consists of two elements:

- Desensitisation – is "the scary thing" small enough, low level enough, far enough away?
- Counterconditioning – adding something nice to the mix

In humans this is usually accompanied by being put into a relaxed or mindful state by a therapist. We can't induce this immediately in horses in the same way, which is why meeting the horse's ethological needs is so important, as well as using enrichment activities to promote positive mood and relaxation. These are what we attended to in our earlier discussions of treatment planning.

In desensitisation we need to pay attention to the horse's coping threshold – where is the point that they are paying attention to the stimulus? Where are they noticing it but not showing a fear response to it or any other indication of an emotional response that isn't one of calm relaxation? We describe this as "attention without tension".

Desensitisation is often done in practice by moving further away (once you are at the "noticing" point) and then reapproaching. Or moving the trigger (the problem stimulus) away and then re-presenting it. Over time and repeated presentations, this shifts the horse's emotional threshold and the reactive response reduces. Be sure to avoid flooding; the horse needs to be at a distance where no escape or avoidance motivation is triggered, so if you have a horse in a stable or on a lead rope while you are presenting a trigger, and there is lead rope tension, or the horse is attempting to leave the stable or head for a corner of it, you are not carrying out desensitisation in a behaviourally appropriate way and you have moved into the domain of flooding. We have discussed the ethics of flooding in earlier chapters, including the safety and emotional risks of forcing a horse to "face their fears" in this way.

In counterconditioning, the objective is to stay at the point of "attention without tension" (noticing the stimulus without showing fear, anxiety or stress) and then to add in a pleasant stimulus (typically food) at that point. The pleasurable response to the food is used as a pathway to changing the negative conditioned emotional response to the problematic stimulus which over time changes the horse's response from fearful (or angry, frustrated or confused) to calm and relaxed. When we're talking about counterconditioning, the "counter" part means to do something differently and the "conditioning" element relates to learning a different response.

Gradual exposure treatment in the form of DS/CC can typically involve working in three- to seven-minute sessions, two to four times a day, and progressing in small, carefully planned increments. Some rest days are beneficial too. The treatment must progress in line with the horse's response: the aim of desensitisation is to reduce the over-aroused, emotion-based reaction that a horse has to a specific object or situation, be it other animals (including pigs, cows, donkeys, dogs), vets, farriers, trailers, certain places or events, or certain noises.

Systematic desensitisation is a structured process of gradually exposing a horse to a less intense version of the object or event they are afraid of in such a way that their fear response is not triggered (McLean and Christensen, 2017). It starts by exposing a horse to a weak, less threatening version of whatever it is they fear or dislike. The object or event can be weakened by making it smaller, slower, shorter lasting, further away or less noisy or being still rather than

Figure 10.1 Desensitisation – letting the horse choose the distance. Equine podiatrist Bonny Mealand works with an inquisitive but cautious Takhi. Photograph by Jean Sinclair.

moving. It can also be "taken apart" into its component parts and systematic DS/CC applied to each element in turn before putting the previously feared object or situation back together again. Over time, as the horse finds they can calmly accept the situation at that low exposure, the stimulus (person, animal, place, object, noise, event, etc.) can be gradually made stronger again by, for example, bringing it closer, increasing its volume or having it move.

A systematic desensitisation plan starts with exposure to a version of the feared thing that the horse can cope with, without being triggered, and gradually over time moves to stronger versions until full or normal exposure is reached (Gough, 1999).

Classical counterconditioning means setting up the situation so the horse over time develops a pleasant feeling and response towards the thing that they are afraid of. The aim is to change a fear response to one of pleasure (via food or scratches) or fun. We do this by associating the feared thing with something good so that it predicts good things for the horse. As soon as the horse sees the scary object (from a distance where they do not respond with fear or anxiety), they are given a treat (or scratch) to elicit a pleasant emotional response. By repeating this many times, the horse learns that whenever the scary thing appears (at a "strength" that the horse can cope with without showing a fear reaction), nice things appear. Eventually, this process develops a positive emotional response to the sight of the previously feared object.

Using appetitive stimuli – good things

If food rewards are being used it's important to use ones that the horse values (i.e. they enjoy), but not too much. If the food is too highly valued there is a risk of frustration during the withholding phase ("how do I get hold of that treat!?") and over-arousal. Some horses can be fed a handful of nuts as a treat and they would be fine. Others, who find this a high value reward, might do whatever they could to get the handler to feed them that treat – now! For those horses who value food very highly, you might choose to use chop, grass nuts or hay cobs, or even carry a large bag over your shoulder containing hay or dot bucket feeding stations around so that when you want to give the horse a treat you walk to the feeding station and drop the treat in the bucket placed there. Alternatively, if you have a horse who loves a scratch you may not even need to use food as a reward, a scratch may be sufficient. The other option, if you have a horse that is particularly pushy or bargy around food, is to use desensitisation alone, without any counterconditioning food during the exercise (within the boundaries of the horse showing no anxiety) and at the end of the session provide some food either in a bucket or scattered on the ground. This has the added benefit of signalling to the horse that the session is finished, and avoids the horse learning, unhelpfully, that there is always a chance of treats in your pocket and seeing you as a permanent treat dispenser. This is also an example of the Premack principle in operation, which you learned about in chapter 4. Using protected contact (see later) can also be useful, where you stay on the other side of a barrier from the horse, protecting you from any bites or kicks and protecting the horse from over-encroachment by the practitioner.

Rewards that mean something to a horse can be:

- a tasty treat (but not so tasty that they get frustrated and forget what you are asking them to do)
- a small feed of chaff with tasty carrots, apples or other favourite fruit and veg chopped up
- scratches (usually on neck or withers, but sometimes shoulders, jaw or chest)
- opportunity to eat grass or browse on hedgerows
- time spent with an equine friend

If using hand-fed food rewards, there is a risk that the horse becomes a bit over-eager and starts to "mug" you. Options to reduce this risk are:

- have treats in a "bum bag" that only appears for the short training session then disappears again
- when feeding the treat always hold your hand so that the horse's head and neck are straight in front, not turned towards you.
- have a bucket and throw the treat into it, so that it appears "as if by magic" in the bucket and not from your hand
- give great scratches instead!

If the horse finds neck scratches rewarding (you see their nose and lips quivering) you can give a neck scratch as a reward. Alternatively use low-value treats such as chaff or hay but don't just give one, give a few so that the horse can chew for a short period of time. This promotes salivation and increases the likelihood of the horse viewing this experience as rewarding. Keeping the sessions short and giving food at the end of the session can be more effective than constantly giving a treat after every try which can create confusion or frustration if the horse becomes more focused on the food than anything else.

Operant counterconditioning can also be used to train a horse to do something different, as an alternative to an unwanted behaviour. In training parlance this is called differential reinforcement of an alternative or incompatible behaviour (DRA/DRI – something the horse can't do at the same time as the unwanted behaviour), and it is part of the operant conditioning paradigm; it's operant counterconditioning, involving a consequence, as opposed to the classical conditioning, involving pairing stimuli to change an emotional response, which we have been discussing until now. Such substitution of one behaviour for another can be achieved by ensuring that the horse is taught the operant association between a new behaviour and the reward in a neutral situation first of all (i.e. one that has nothing to do with the context in which the problem behaviour occurred), and then move this behavioural response over to the original context. Meanwhile, the unwanted behavioural response in the first context needs to have been prevented.

Consider the example of Bertie who frequently paws the ground when tied up outside the stable (often met with a shout of "oi!" from somewhere on the yard, with no reducing effect on Bertie's behaviour either because he has habituated to "oi!" or it is in fact positively reinforcing his behaviour by providing attention):

1. prevent the original response – don't tie up outside the stable (this facilitates extinction of the unwanted response – pawing the ground)
2. train an alternative/incompatible response – tie up somewhere else, and reward for all four feet on the floor (Bertie cannot paw the ground if he has all four feet on the floor)
3. substitute the old response in the old context – move back to tying up outside the stable
4. reinforce all four feet on the floor in this context (and ignore any unwanted appearances of the old pawing behaviour)
5. over time you'll be able to fade out the reinforcement as Bertie's replacement response becomes embedded, but remember the rules of extinction; if Bertie never experiences any consequence for the all four feet on the floor behaviour ever again, the replacement response will extinguish, so remember to periodically reinforce this behaviour.

This "remember to reinforce" principle applies to all our interactions with horses; it pays to get into the habit of being aware of what a horse is doing and remembering to reward anything that you like, from time to time, with food if available at that moment or with a scratch.

Something to be aware of in using operant counterconditioning as behaviour modification though: hopefully the thought has probably already occurred to you: why was Bertie pawing the ground in the first place? This is the direction that I prefer to approach unwanted behaviour from, first understanding the reasons why a behaviour emerged, and resolving any deficits that contributed to its development, before moving on to training an alternative behaviour. In fact, as you'll see in other examples in this book, by understanding the horse's emotional motivation and by providing what the horse needs in order to feel relaxed and resilient, it's possible to resolve many unwanted behaviours "from the bottom up", such that the top-level behaviour that was causing the owner the problem either stops appearing or becomes a lot easier to retrain through differential reinforcement.

Gradual exposure therapy can be used with anything that is causing a difficulty for the horse, for instance:

1. Leaving other horses (gradual exposure to distance and duration of time away from them)
   - Can the horse walk 100 m from their friend/herd and remain calm?
   - Can the horse walk all the way to the gate and remain calm?
   - Can the horse walk through the gate and a few metres down the path and remain calm?
2. Tack (gradual exposure to seeing tack, tack being presented and put on)
   - Can the horse remain calm when they see the tack from 5 m away?
   - Can the horse remain calm when the tack is held up to their side?
   - Can the horse remain calm when they are approached with the tack whilst eating hay?

3. Grooming (gradual exposure to being stroked, rubbed, scratched, sight of a brush, brush used very briefly on a part of the body the horse can easily accept)
   - Can the horse stay calm while being stroked?
   - Can the horse stay calm while their lower neck is brushed?
   - Can the horse stay calm when brushed gently all over with a body brush?

Note that these are examples for illustrative purposes; real-life cases should all be properly assessed and treated individually and are likely to include more detailed steps than the prior examples.

The first essential task for the behaviourist is to find out where the horse's threshold is. How close are they before they back away, stop, not want to go closer or look worried/frightened? The point where the horse starts to look uncomfortable is that horse's fear or anxiety threshold. The aim is to change that threshold (but we don't start our DS/CC from this point! We start further back; remember, attention without tension.) The higher the threshold (imagine a wall) the harder it is to breach, the more relaxed the horse will be in the presence of the (formerly) scary thing and the more resilience they will develop. The "threshold" that we discuss here represents a set of processes and structures in the horse's brain, which is why if the threshold is going to change, we are always operating in the neurobiological realm of influencing changes in the horse's neurocircuitry.

Thresholds

You can observe body language to tell if any horse has gone over their "coping threshold" and is likely to be experiencing anxiety:

- raised head carriage – often associated with tighter neck musculature
- reluctance to go forward – slowing down or planting
- horse "grows" and appears to get bigger
- triangulation of the eye – the eye shape changes from soft and round to looking as if the front of the top line has become sharp and triangular in shape
- tension in the mouth and chin – the muzzle changes from soft and round to looking angular and "squared off" and you can see a distinct, tight chin groove

Guidelines for a successful DS/CC BMP:

- Bring the scary thing into view (by moving it nearer to the horse or moving the horse nearer to the thing).
- Give small treats as long as the thing is in sight.
- Move it out of sight or move the horse away.
- Stop the treats as soon as the scary thing is out of sight.
- Repeat this pattern of "approach – give treat (or scratches) – retreat – stop treats" at least ten times each session (sessions will usually be four to seven minutes long).
- End each session on a good note, when the horse is really soft and relaxed, isn't at all worried and has stopped even noticing the feared thing.
- Start each new session at the same level of exposure (or ideally a little below so there is an overlap) that you ended on in your last session. Stay at this level as long as the horse needs, to experience the scary thing without tension and be relaxed before moving on. This may take several sessions across several days.

- Do not move on to the next level of exposure until the horse shows no fear or anxiety at the start of a session and is relaxed and calmly taking food (if used) at that exposure level.
- Never start a new session at a higher exposure level than you ended the last session on.
- End each session with a "jackpot" treat, something different and more interesting to the horse than the treats used in the session.
- Avoid frightening the horse. Make sure they are presented with a version of the scary thing that they can cope with, without showing any fear.
- If the horse does show fear, increase the distance or reduce intensity in some other way such as by reducing its sound or movement.
- Where possible avoid the horse experiencing the scary thing outside the treatment sessions. Ideally the horse should only be exposed during treatment, when you can control the situation and you can use counterconditioning with treats or scratches.
- Once the horse can be fully relaxed in the presence of the normal version of the object they used to be scared of, keep up maintenance sessions two or three times a month to maintain this new behavioural response.

In reality gradual exposure means changing the criteria – distance, duration, frequency, context or intensity – one element at a time. Start below threshold and gradually expose in manageable increments. Distance might mean enough distance to ensure the horse doesn't feel restricted or pressured. Duration might be initially starting at ten seconds before gradually building up the time over time. Frequency might mean starting with four repetitions, and context might mean changing where you do the training. Intensity: at all times be aware of the horse's emotional response as an indicator of whether to proceed or to go back a few steps.

Common themes in behaviour consultations

With the exceptions of extreme aggressive behaviour, loading and separation anxiety, the majority of my cases fall under the somewhat messy description of "unhandleability" – whether ridden, on the ground or for routine vet/farrier/health checks and treatment. Examples include:

- explosive behaviour (bucking, rearing, bolting, spinning)
- napping, planting/stopping
- not being able to be led into or out of the stable
- nipping/biting the owner/handler
- running round on the end of a lead rope
- unable to be tied up or constant walking from one end of the lead rope to the other whilst tied up

The BMP (behaviour modification plan/programme; you may also hear this referred to as a training or a shaping plan) is like a ladder, identifying where are you now and where you want to be. For trailer loading, the top of the ladder, the end goal, might be the ability to load the horse into the trailer, tie them up, move the partition back into position, fasten the breeching strap, close the ramp, drive to do the activity, unload and repeat the process to come home, all done in a calm and relaxed manner. That's the end point, or the goal, and the shaping plan provides the steps to get from the starting point (the place where the horse stops looking comfortable) to the end point. It's a bit like building muscle in the gym. The muscle needs to be conditioned in such a way that it adapts and becomes stronger, but not so much that it fails and breaks down or becomes injured. It's the same in terms of conditioning the brain; the steps mustn't be so big that they cause too much stress and damage, but equally they mustn't be too small with insufficient effect. This process is not a linear one, and it's important to recognise when the horse is ready to move to the next step and when they might need to go back a step. To that extent there is an element of a step by step process which may sound a bit formulaic, but actually it is the responsibility of the behaviourist and the owner to apply their own awareness and judgment in these moments – what is the horse's body language telling me? Have they been completely and consistently calm? If the decision is that the horse is ready to move up a step then consideration needs to be given to how far to move up and what the best next stage is. There is risk in focusing solely on doing a step of the training plan for exactly *x* number of days and then moving to the next level without actually checking in with the horse and their threshold levels. It is not a failure to have to repeat a step or go back down a rung or two – that's normal training practice. In fact it would be surprising if there wasn't the need to do that at some point during the BMP. If the horse is exhibiting signs of being close to threshold – signs of tension, tightness in mouth or upper eyelid, appearing to have grown, moving a bit more slowly or some stickiness, then it is time to reassess. Is the horse ready to move forward? Should I stay at this step for a bit longer? Or have I pushed it a bit far and need to step back a bit? It might be that you rearrange the order of the steps or decide that some of the later steps are actually optional and not a requirement to achieve the owner's stated goal. This isn't a one size fits all approach, it depends on the horse, the facilities and the resources the owner has. As a behaviourist it is our job to assess, given these factors, what is likely to work best and if something isn't working to try a different approach. You need a variety of tools/techniques in your behaviour modification toolbox and you need to be prepared to take the time it takes. That might be three weeks, it might be three months or it might be three years. It is the horse who shows when they are ready to move to the next step.

When reading the sections of the following BMPs, remember that by now we are talking about the third stage in a treatment plan that has initially met the horse's needs for safety, companionship and species-specific behavioural expression (enrichment). If you feel that the horse you are treating is not showing a regular relaxed outlook in their daily life, they may not yet be ready to move on to this stage, and you may need to review and repeat earlier stages, or parts of them. Always have in your mind the interactions between health and behaviour; does the vet need to be involved to assess pain or discomfort at any time? Try also to think about how each of these behaviour problems, as perceived from the human side, are actually natural behavioural responses for horses, in terms of their evolution and their ethogram.

Separation anxiety

Horses feel safe on their familiar home range and with a consistent, unchanging group of horses of their choice. Understanding this, it's easy to see how challenging it can be when we ask them to hack out alone, stay in while their companion is turned out, travel alone and enter a competition arena while their travelling companion is warming up elsewhere.

As ever, the treatment for separation anxiety needs to be done in very small, step by step stages because modification of the behaviour comes from changes in how the horse's brain processes information, changing their negative associations, when separated, to positive associations. The action plan involves breaking down separation into smaller components and changing the horse's emotional associations with each component part so that gradually they become confident to be separated in different situations and with different horses. The component parts might include:

1. taking horse 1 away from horse 2 in the field
2. taking horse 2 away from horse 1 in the field
3. nature walks increasing distance between horse 1 and horse 2
4. taking horse 1 away from horse 2 in the stable
5. taking horse 2 away from horse 1 in the stable

6. "wavy" walking with both horses (one person walks each horse in parallel with the other, sometimes moving further apart and sometimes closer together)
7. taking horse 1 away from horse 2 in the arena
8. taking horse 2 away from horse 1 in the arena
9. visiting time for both horses (where they are stabled separately, or kept in separate paddocks, for a period of time, and one horse comes to visit the other over the fence or the stable door)
10. leap frog for both horses (walking or riding both horses, each takes turn to be in the lead and moves further away from the one behind)
11. including a different horse in the training to learn to be comfortable when separated

If there are any signs that the horse is close to or over their coping threshold at any stage, then return the horse who is being removed immediately and try again from an earlier step where they can remain calm. Remember it's not at all helpful to reach this over-threshold point regularly; if this is happening, rethink the BMP, either backstepping to a much earlier stage or adding more and smaller steps into the plan.

Touch acceptance

Horses enjoy human touch in the form of scratches and pleasurable stroking, but many horses become sensitive to touch; no surprise when we consider how forceful weaning and a "no choice" approach to introducing grooming is likely to have flooded them into anxious acceptance as opposed to a relaxed outlook. Many owners don't realise that their horse is experiencing any touch sensitivity, because they always handle them with restraint (tied up) or in a small area of no escape (the stable). A choice test can reveal interesting and enlightening findings. When horses are struggling with other areas of their life and are generally closer to their coping threshold on a regular basis, touch sensitivity can emerge in the form of fidgeting, nipping, even seemingly unrelated behaviours such as refusal to be caught (what predictable happens next? Usually grooming – unpleasant touch). For this reason I like to include touch acceptance as a transition BMP towards the end of the gaining resilience stage of a treatment plan.

A touch acceptance BMP means working through a programme where the horse learns that they can stay calm while their owner reaches out and touches various parts of their body, starting at easy access no/low reaction places and working through other areas in order of difficulty. Horses have different touch sensitivities to different parts of their body, for example their shoulder may be less vulnerable, feet are more vulnerable, although always be aware of individual differences. Use the principles of desensitisation and counterconditioning described earlier to gradually move from stroking "easy" places to harder ones. You can do it without food present, if the horse is unable to remain calm when food rewards are used, or you can give food each time you touch the horse, or you can stroke for a short period and give food in a bucket at the end of the session. This helps to replace the automatic fear and conflict response with a pleasant and relaxed emotional response, and has the added advantage of enhancing the relationship between the owner and the horse.

Calm lead rope walking

For this you can use scratches or low-value food treats, a very long lead rope and working behind a fence or other barrier (protected contact) if safety is a concern (however as ever, if you think safety is a concern – why is this? Perhaps you need to go back some stages in the treatment plan to resolve this at an earlier stage first).

You may wish to start by training the horse to touch and follow a target so that they learn to walk with a person without needing a headcollar (see the target game at Appendix 5). A headcollar is another example where owners often don't know what their horse would do without one; it's easy for this ubiquitous piece of equipment to become either a discriminative or conditioned stimulus whereby the horse learns "there's no point in me trying to express any of my own preferred behaviour here" (think back to the concept of learned helplessness when we discussed learning theory). Objective achieved, from the owner's perspective, they want an obliging horse, but there is an emotional cost to the horse of learning a predictable, "what happens next", when that is a predictable loss of the opportunity to express choice and agency.

When you have progressed to using a headcollar and lead rope, you can train the horse to give to a very light pressure of the lead rope in order to help them learn to respond to light pressure steering without fear or anxiety.

The next step might be to take the horse on nature rambles to explore and investigate their environment, using neck and wither scratches as a reward. Use a very long lead rope, or a lunge line, and walk in hand around the yard or down a lane, letting the horse stop, sniff and explore things. The aim is for the owner to take an interest in the same things, as if they are exploring together (they may need to feign interest, as if going for a walk with a toddler!). This helps to develop a good relationship between horse and owner and raise the owner's awareness of body language and natural exploratory behaviours.

Using the mounting block

As ever, consider why a horse might develop an aversion to a mounting block and address those factors first. This BMP begins with finding the threshold for approaching the mounting block – how far away does the horse need to be to show no anxiety? From that place start gradual exposure by approaching, retreating and repeating (this is the desensitisation in DS/CC). Make the gradually increasing proximity of the mounting block a nice experience (this is the counterconditioning in DS/CC) by giving a food reward, scratches or a stroke. Both food and pleasant touch are known as "biologically important stimuli" that promote pleasant mental and emotional experiences for the horse. Then add in a rider stepping onto the mounting block and incrementally add in:

- picking up the reins
- rider lifting foot
- left foot into the stirrup
- weight in stirrup
- all weight off right leg
- lift right leg part way up
- lift right leg all the way over
- sit in the saddle
- ask horse to move off

Start and end these sessions with a consistent start and finish signal; one to start it and one to end it, that are not used elsewhere in interactions with the horse. Typically I start by saying "training time", and end a session by holding my hands up, palms outwards, saying "finished" and scattering some low-value food on the ground. Predictability in life helps to promote relaxation and resilience. Remember, horses are "hard wired" through their survival instinct to always want to know what happens next and the more we can help them with this, the less stress they will experience.

See also Appendix 11 for an example shaping plan for DS/CC to the saddle.

Getting comfortable with being tied up

It's common for horses to learn that it's ok to be tied up in a couple of places, such as just outside their stable or outside their trailer, and often owners don't learn this until they try to tie their horse up somewhere else and find they've pulled back and broken the baling twine as soon as their back is turned. This demonstrates the importance of context-specific learning in horses and the need to "proof" any learning for generalisation. For this it's important to recognise that the horse may not be able to be tied up for as long in an unfamiliar place (stable or part of the yard) so expect less, and in later stages repeat the following process in a number of different locations. Firstly, feed the lead rope through a tie ring (or sometimes through the stable bars as an earlier step) and have a helper holding the end of it (the other end is attached to the headcollar). This allows the horse some room to move around without feeling the sharp pull when they reach the end of the lead rope. They can then gradually get used to the pressure rather than it being a sudden tug if they pull back. This may not be appropriate for all horses, particularly if the presence of a number of people is likely to worry the individual horse. As an alternative, person 1 could feed the lead rope through a tie ring or bar themselves and hold it, giving scratches or food. As the horse becomes accustomed to this experience in an unfamiliar situation, start to add criteria, always one at a time and always going back a step or two when raising criteria. For example:

add duration – tied up (or held, lead rope fed through) for longer
add distraction – handler does some grooming or picks feet out
add difficulty – start to tie to a ring, always using baling twine or some type of fastening that will easily break if the horse pulls back

Learning to stand still

A learning to stand BMP involves using scratches or a food reward for standing and then allowing the horse to walk around if they want to (relief from standing still). Make sure you let the horse walk round before they get to the stage when they take matters into their own hands (or hooves!) by fidgeting or walking off. Rope wiggling isn't helpful and may add to anxiety depending on any previous training the horse has experienced. Standing for a long period of time when it's not the horse's choice is not easy for many horses (remember their evolution) and needs to be introduced gradually at their own pace whilst avoiding situations that may cause them stress.

Learning to be comfortable about being caught

This BMP might include some or all of the following steps, each repeated several times:

- go as close as you can to the horse without seeing any tension, and put a handful of chaff (can be mixed with fruit, vegetables or a few nuts)
- don't keep trying to catch an avoidant horse. At the same time, you don't need to completely leave the field, instead wait at a short distance in the field and as soon as the horse turns their head to you, say "good" and toss food onto the ground, then you can leave the field.
- a few times a day, go into the field, give the horse some chaff mix and then leave again. This helps to break up the association between being caught and "something" (often unwanted) happening next.
- repeat, carrying a headcollar and lead rope
- put headcollar on, give food, take headcollar off, walk away

- repeat but lead horse a short distance then take headcollar off and walk away
- build up leading until horse is accompanying you or owner to the yard
- repeat and take horse into stable, ensuring there is a bucket with some chaff mix waiting for them in there

If you leave the stable/field as soon as the horse has avoided being caught, they may learn that in order to get rid of you all they have to do is the "not being caught" behaviour and this may cause that behaviour to be repeated. Instead, if you give a treat and leave the stable/field when they show the tiniest piece of interest in you, you are "ending on a good note" and setting yourself up for a better possibility of success next time you try and catch them.

DS or CC?

At this point it's worth mentioning that a BMP does not always have to be desensitisation and counterconditioning combined. In the previous example, a first stage of learning to be caught might just be desensitisation – gradual exposure at the "attention without tension" distance – before the handler could get close enough to start to add food (counterconditioning) (Figure 10.1). Conversely, in the following example, feeling comfortable about the headcollar, consider a horse on box rest who has developed a conditioned emotional response to the headcollar; there is no opportunity to take the headcollar far enough away, you are restricted by the stable walls. In this situation, a classical counterconditioning protocol where the horse is presented with food when the headcollar is visible, and the food presentation stops when the headcollar goes out of sight, can be effective if carried out consistently.

Learning to be comfortable with a headcollar

Use start and stop signals (e.g. saying "training time" to start/"finished" to end) to indicate that the session is about to start and when it has finished. The first step is to show the horse the headcollar at a distance where they do not show any avoidant behaviour, toss food in a bucket that is further away (so the horse doesn't need to come closer to the headcollar for the food) and then, importantly, move *further away* with the headcollar. Do this approach and retreat type of process about seven times and then stop (this is one set with seven repetitions) and give the finish signal. Ideally then put the headcollar out of sight; you might need to cover it up with something or hide it. Certainly do not place it closer to the horse than the distance you have been working with. This process can be repeated up to four times per day and you will need to leave as least 30 minutes between sets.

It's not uncommon to need to use the headcollar outside these training sessions for hoof trimming or vet visits and so on, and this could risk undoing all the good DS/CC work done thus far. Helpful options here are to use a different headcollar, e.g. swap a fabric headcollar to a leather one (I don't recommend using rope halters), and ensure hay, haylage or plenty of chaff is available for the horse to eat during the procedure.

Suggested steps to help a horse learn to stand calmly when the headcollar is presented:

1. have the horse behind a fence with you on the other side
2. check that you are at the appropriate "attention without tension" distance. The "right" distance at this time is where the headcollar can be seen but the horse doesn't react to it, doesn't look away, put ears back, yawn or lick/chew.
3. say "training time" (the discriminative stimulus signal that gives clear predictability about what will happen next and the opportunity to obtain rewards in this context)

4. take one step towards the horse, holding the headcollar in front of you for three seconds
5. do not let the headcollar go over the fence onto the horse's side
6. scatter chaff mix to add a positive association
7. take one step back and put the headcollar behind your back
8. repeat this about seven times
9. put headcollar out of sight and say "finished" – hold empty palms out for the horse to see (your end of session signal)

Progress to being able to hold the headcollar closer to the horse with no reaction, and repeat the aforementioned steps. Remember you can use a marker too if you wish: when you hold the headcollar up to the horse, say "yes", "good" or use a tongue click (decide on one of these options and be consistent) before giving a handful of food that encourages the horse to chew. The "yes", "good" or tongue click is the marker and functions in the same way as a clicker (see chapter 4); it marks the behaviour you want and tells the horse a treat is coming. In this case the marked behaviour is standing calmly while the headcollar is presented. Once the horse is calm and shows no reaction for up to six of the seven repetitions you can move on; repeat the steps but this time with the headcollar open as if you were going to put it on. Mark the calm behaviour and give the food treat. Remember to use the start and finish signals and put the headcollar completely out of sight when you finish.

Protected contact

A quick word about protected contact. Protected contact is a system where the handler and the horse are separated by a physical barrier which enhances both safety and welfare. This can be especially helpful if you are working with a frightened, aggressive or unhandled horse. An important part of protected contact is voluntary cooperation. The horse chooses if and when to participate and is free to move away, creating distance between themselves and the person. This is a good exercise for the human element in respecting the horse's choices and remaining patient if the horse is slow to respond or declines to participate.

Loading

At the stage where a horse is persistently refusing to enter the trailer, their fear response has become automatic and any treatment plan needs to include desensitising and counterconditioning the horse to the trailer. The BMP is based on creating positive associations with the trailer and presenting positive reinforcement (reward) for desired behaviour. This starts with gradually reintroducing the horse to the trailer, starting at a distance where no fear or anxiety is observed and gradually moving closer over a number of short sessions as the fear diminishes.

The horse's fear of the trailer changes to calm behaviour as they begin to associate the trailer with the food/scratch reward and the horse is positively reinforced for approaching the trailer. This is done in short daily sessions, starting at a distance manageable for the horse and gradually getting closer (always at a pace that means the horse can feel calm) until they are calmly touching a target: when they are next to the trailer; then when the ramp is lowered; and eventually following the target to walk up the ramp; remaining calm in the trailer when stationary; unloading calmly; and subsequently repeating this calm behaviour when the trailer is moving. This process needs to be repeated in a variety of contexts to ensure generalisation of the new behaviour.

A target and/or clicker is not always necessary; it depends on the individual horse and with some horses success may be achieved by following the prior protocol and just using food rewards as the horse remains calm and gradually approaches the trailer over a number of sessions. However, other horses do better when being asked to perform the operant task of touching a target, combined with approach and retreat, bearing in mind that as ever, "Pavlov is always on your shoulder", meaning that if you are working through this process in the afore-mentioned way there will be always be a classical counterconditioning effect of pairing the scary thing with the primary reinforcer (food or scratch).

Next are suggested steps to help a horse to learn to load calmly into a trailer, travel calmly to a competition and then travel home calmly (let's say do this procedure for five to ten minutes, two to three times a day, but also work with whatever the client can do on their schedule; you can still be successful doing this two or three times a week).

Pre-training required:

- touch target with nose
- calm back up four steps

Start by finding the horse's threshold. How far away from the trailer or horsebox does the horse need to be in order to remain calm and relaxed? At what distance do they start to notice the trailer without appearing uncomfortable? From this point you can begin the gradual exposure process.

1. "Take the trailer to pieces" as much as you can so that it doesn't look like the trailer the horse usually travels in – take the partition out, remove any back bars or panels. Make sure front and rear ramps are down and there is plenty of daylight showing through the trailer. Don't put in a hay net yet.
2. Take the horse far enough away from the trailer that they are not showing any fear response or anxiety.
3. At points where you see no anxiety and the horse is calm, hold the target out and ask them to touch it with their nose (you have already taught this behaviour in earlier sessions). Click and treat, walk away, repeat by advancing to the same point again (no closer until you have repeated this calm approach and retreat several times).
4. Gradually, and only as long as the horse can remain calm, move closer to the trailer, in very small successive approximations. Work in a ripple/bubble pattern around the trailer and ensure the horse sees it from both sides, that is with both eyes.
5. Raise the criteria:

    - one foot on ramp
    - two feet on ramp and back up
    - three feet on ramp and back up
    - four feet on ramp and back up
    - walk forward on the ramp one step and back up
    - walk up two steps, three steps etc. and back up
    - walk through the trailer and out of the front ramp
    - walk in to trailer, stand for five seconds then walk forward and out
    - walk in and stand ten seconds
    - walk in and out, gradually increase standing duration until the horse can stand for two minutes (target, click and treat during this time)

6. Horse stands in trailer while:

    - your helper walks across and back behind the trailer (ramp is down), then walk them down the front ramp (NB reduce the duration criterion back down from three minutes first off because this is a significant change; then slowly increase the time again)

- repeat while the helper lifts the back ramp just a little
- repeat while they lift back ramp more
- repeat gradually until back ramp is closed
- repeat all previous while front ramp is gradually closed too

7. Start to put the trailer back together with one new thing at a time: hay net, partition, back hinged panel if you decide to use it. Each time you add a new thing go back as far as you need to in the previous steps, to where the horse can stay calm while you build up the criteria again. Build in factors such as gradually fastening the breeching strap and tying up the lead rope inside the trailer.
8. Repeat in various situations where the trailer is parked in different places to generalise the learning. Start to gradually fade the use of the clicker now and then gradually fade the treats, until you are just giving the horse a few treats when they have loaded.

Note that you don't need to do straight line approach and retreat; the objective is to expose the horse to the fear-inducing stimulus without putting them over threshold, and this can be done in a variety of ways, including:

- walk round the trailer at a distance where the horse remains calm and relaxed throughout
- walk in a wiggly line around the vehicle (a bit like a wavy circle) where sometimes you're a bit further away from the trailer and sometimes you're a bit closer, but you never cross the threshold where the horse becomes uncomfortable.

Alternatives to marking and feeding could be to just feed with no marker, or have feed stations at some points on the circle (perhaps 2 o'clock, 6 o'clock and 10 o'clock) where either treats are dropped in or the horse can chew on hay or chaff for a few minutes before moving on to the next feed station. Once you've circled a couple of times, give the finish signal and stop the session before moving well away from the trailer and giving a last handful of treats, bucket of feed or graze.

Once the horse can stand calmly and all of the "trailer experience" has been put back together, you can start the towing vehicle's engine and turn it off again. First start the engine, turn off immediately and unload the horse. Gradually build up the duration the engine runs for. When you have built up the engine running time to five minutes, start to drive very short distances (start with 200 metres) then unload. Gradually increase the distance you drive until you are driving a short circuit out of the yard and back. After this, build up to longer journeys off the yard, then journeys to a nice easy hacking venue (less stressful than a competition). Once you are ready to drive to a different activity (competition/clinic, etc.), aim to get there and just let the horse graze for a period of time before loading them up again and going home. After this the horse can be walked in hand, then ridden around the venue without competing and eventually build up to competing again. Many horses benefit from a travelling companion, if so it can be helpful in early stages if the companion remains within sight during the time the horse is warming up and ideally when in the arena (competing hors concours might be a helpful step to include here).

Remember, each of the BMPs suggested is an example. The specific BMPs that you will develop as a behaviourist will be part of your bespoke treatment plan, itself a product of your behavioural assessment, the horse's temperament, their environment, the resources (including time) available to their caregiver, the caregiver's skillset and commitment, and of course you will review progress and adjust aspects of the treatment plan and specific behaviour modification programmes as you continue to work with your client, bringing us on to a discussion of communicating with your client in the next chapter.

Figure 10.2 Team work, the end result. Enjoying a pleasure ride as a partnership. Photograph by Bryan Roberts.

Figure 10.3 Team work, the end result. Sam Campling on Scarlett Rose at Somerford Park horse trials, 2019. Photograph by Matt Nuttall Photography.

References

Gough, M. R. (1999). A note on the use of behavioural modification to aid clipping ponies. *Applied Animal Behaviour Science*, *63*(2), 171–175.

Hothersall, B., & Casey, R. (2012). Undesired behaviour in horses: A review of their development, prevention, management and association with welfare. *Equine Veterinary Education*, *24*(9), 479–485.

McGreevy, P. (2012). *Equine Behavior: A Guide for Veterinarians and Equine Scientists*. Elsevier Health Sciences.

McLean, A. N., & Christensen, J. W. (2017). The application of learning theory in horse training. *Applied Animal Behaviour Science*, *190*, 18–27.

Mills, D. S., & Nankervis, K. J. (2013). *Equine Behaviour: Principles and Practice*. John Wiley & Sons.

Rogers, S. (Ed.). (2018). *Equine Behaviour in Mind: Applying Behavioural Science to the Way We Keep, Work and Care for Horses*. 5m Books Ltd.

Zeitler-Feicht, M. H. (2004). *Horse Behaviour Explained: Origin, Treatment and Prevention of Problems*. London: Manson Publishing.

Chapter 11

Communication skills in behaviour consulting

Introduction to human behaviour change

Now we turn our attention to the human side of the relationship: how to conduct a behaviour consultation with a horse owner in order to elicit the information required to inform the case history and treatment plan and to educate, support and coach the owner as they work through the plan you have developed for them. This includes considering the relational aspects of communicating, which is particularly important given that human behaviour change is fundamental to successfully modifying the horse's behaviour. At the first meeting it is important to establish a relationship in which the client has confidence in you and your advice. The aim is always to empower the client to understand and define the problem they are experiencing in order to set goals and to take action. The client must not feel judged but rather enabled and empowered to communicate freely with you (Daniels *et al.*, 2023). It is important that as part of this you attend to the client's aims in requesting your help by asking them what a successful outcome would look like for them.

Conducting a behaviour consultation

On the human side of a behaviour consultation, the effectiveness of your assessment and treatment plan lies in large part in your ability to obtain the information you need and to educate, support and coach the owner as they work through the plan you have developed for them. I base my behaviour consultations on communication models found within Gerard Egan's skilled helper (Egan, 1990) and Eric Berne's Transactional Analysis (Stewart and Joines, 1987), as well as Carl Rogers' person-centred approach (Cooper, 2024).

In general, conducting a behaviour consultation is a three-stage process: hearing your client's story; clarifying their goals; agreeing the action they will take. The first stage is to let your client tell their story. Through this narrative we can find out what is happening in the life of both client and horse, start to identify what might be missing or what might need to change and where the owner may be missing opportunities to help their horse. Exploring the "problem" as the client sees it also helps to build rapport and allows us some insight into what the client

DOI: 10.1201/9781003581390-15

might not be observing. Often they don't see the anxiety in their horse (it's just how she is) or the signs of pain (he always does that when I do this). Sometimes their interpretation is that the horse is being "naughty" or disrespecting them and sometimes that's because it's what they've been told by the horse people they spend time with. Many clients don't realise that their horse's behaviour is underpinned by emotions (just like us), and once they realise this, they are able to be more empathetic.

It's important to let your client talk without jumping in, without interpreting what they're saying or explaining why they might be right or wrong. The main skill here is active listening. The ability to be attentive and demonstrate your empathic attention through non-verbal signs such as soft eye contact (no staring), nodding and perhaps the odd "uh-huh" is something you can practice. Give your client space to talk by using silence to allow them to gather their thoughts and if they appear stuck about what to say next, reflect back what they have just told you. Use questions sparingly and just to elicit further information. Open questions are always preferable to closed questions which only require a yes or no answer. A useful rule of thumb is to focus your questions on establishing when, what, where, who and how.

The second stage in the process is understanding the client's thoughts and feelings and what a successful outcome would look like to them. Stage one sets the scene and when done well will have built rapport and established trust between you and your client. You should now be in a position to assess how committed the owner is to helping their horse and working with you to do this. At this point some decision-making is required on both sides in choosing what goals to set. Ask the client what the essentials (needs) are and what the desirables (wants) are. How much time do they have to commit to any behavioural rehabilitation work? Are they prepared to do the work you suggest? You can elicit this information in one of three ways:

- by asking direct questions
- by tentatively talking around the subject of other commitments, limitations and barriers (time pressures, attitudes of family, friends, other liveries)
- by inferring their commitment from the "explore and understand" work you have already done

During this second stage your questions can become more closed, allowing you to further probe for information and to challenge, in an appropriate way, some of the client's perspectives, beliefs and values. This is the time to help your client see things in a different way, in counselling we call this reframing. For example, helping them recognise that their horse is frightened not stubborn or that their horse has emotions and the capacity to make choices and to learn from those choices within the context of their own species. Start to develop a vision together – what the client wants to be able to do – help them imagine it and visualise doing it. To do this it's helpful to:

- use probing and challenging questions to identify what the possible options are for this client in terms of a pathway to a successful outcome
- agree (co-create) outcomes and goals with the client; although you are the behaviour expert, they know their horse and you both have essential qualities to bring to the table
- test commitment and then think about how you can develop that commitment by using your communication skills and your relationality – giving encouraging feedback and setting time-based milestones that are easy for the client to commit to

Goal-setting questions

What does the client need?

What if that need doesn't match what their horse needs? (Do you see your role as to negotiate on behalf of the horse?)

Will the client give up some wants in return for simplicity, or are they prepared to compromise or prioritise?

Do the solutions make sense for the client as well as being in line with what the horse needs in terms of their species needs and treatment needs?

Will the client "sign up" to these goals? Remember that this is an iterative process and the goals are not set in stone. The process can be reviewed at any time.

Does the client have the resources needed to reach the goal – time, environment, support, mindset?

How can the client make better choices?

Is the client choosing freely? Do they believe in their own agency and ability to change?

What is the client prepared to pay or give up, either permanently or temporarily? For example – changing their dressage trainer to one with a better understanding of horses and learning theory, not travelling to horse shows, no riding for a period of time, horse not being as well groomed as usual.

Client example – helping the client to identify and describe a successful outcome:

- Willow shows soft, friendly body language
- I'm able to lead, groom and exercise her without any presentations of aggressive behaviour
- Other people on the yard are able to bring her in, feed, muck out and turn her out without being fearful of any aggressive behaviour

By this time, the discussion is likely to include talking about the reasonableness of the goals and the client's expectations. Equally, what education does the owner need to enable them to make informed and realistic decisions about their goals? In essence, our role as behaviour practitioners is to support our clients to be the behaviour therapist for their horse. We're almost working in a "one removed" way to help the client help their horse and more often than not that requires behaviour/mindset change from the owner too. It's helpful for them to understand why they're doing what they're doing. In the end, the decisions about what training they're going to do, what activities they will undertake, where the horse is going to live, if they're going to sell the horse, if they're going to euthanise, all rest with them. Our job is to help set them up to make informed decisions.

The third stage of a consultation is about action planning in ways that are meaningful to the client. In order for the treatment plan to be successfully implemented it needs full involvement from the client in identifying ways to achieve their goals as well as being congruent with their values and beliefs. The aim is to set actions that are achievable and realistic for the client in front of you at that moment. In essence, meeting the client where they are, not where you think they should be. Time management is crucial; if you don't set milestones with dates then you'll make little progress or slide to a halt. I generally do this by arranging follow-up times with clients and asking them to update me on their progress with the tasks on their action plan. Ongoing evaluation is essential, through follow-ups and the client's own observation and recording of changes – possibly using progress charts or whatever works best for the client – smart phone, notebook, chart or suitable app.

The action plan needs to identify possible strategies, and this is where prompting comes in using the information you have acquired throughout the consultation. You can prompt with suggestions and check them out with the client. For example they currently feed with a hay net, could they spread hay on the stable floor to simulate natural "eat and move" behaviour? Make sure you link the treatment plan to both the client's goals and meeting the horse's needs in terms of physical and mental health. This might be a good time to refer back to the five domains model covered in the welfare and wellbeing chapter.

When setting the action plan it is important to be clear about:

- what resources are available?
- how it will fit into the horse/human's current lifestyle?
- what skills are required? (training/handling)
- what knowledge and understanding is required? (it is our role to help educate the client)
- who else needs to help them? (vet, physio, other paraprofessionals, family, groom, livery yard owner/manager, etc.)
- how will the client know where they are now and where they want to be? how are you going to measure progress?
- how will the client know when they've arrived at their goal and is this still where they want to be?

In short, these are the steps you'll go through with your client in a behaviour consultation:

- let the client tell their story
- tease out the actual problem
- set overall aim/goals
- split into milestones
- monitor progress on the road to each milestone
- review and adjust as necessary
- know when the goal has been achieved
- celebrate success

Success depends on the quality of your communication

Research in both human and horse behaviour change programmes shows that practitioners who collaborate with their client in the development of the treatment plan create a more positive experience and as a result achieve greater adherence to the plan and subsequent improvement in the unwanted behaviour (HBCA, 2019; Wolframm *et al.*, 2023). This client-centred approach promotes an empathic, nurturing rather than authoritarian communication strategy where the client is an equal partner in the process. Its main tenet is building empathy with the client by gently exploring options without being judgmental or critical of their current beliefs. The behaviourist's role is to explain the likely cause(s) of unwanted behaviour, co-create strategies to help resolve it and provide a treatment plan that is both bespoke and flexible.

Motivational interviewing is another important concept in client-centred communication which requires you to use open questions and active listening in order to evoke and take advantage of your client's motivation for behaviour change. Such information gathering is facilitated through relaxed and open interviewing that helps the client to understand the entire treatment process. For example, allowing clients to develop/co-create their own environmental enrichments ideas (rather than telling them what to do) means you are providing them with the autonomy and choice that is positively correlated with successful treatment outcomes. The other key component is providing the client with the opportunity to question and explore your advice in order to fully understand the problem and support the actions that you propose. This ensures the treatment plan is a tailored, flexible and incremental set of actions which makes sense to the client and is therefore more likely to result in compliance over time. Being realistic about expectations of behaviour prognosis is also linked to client engagement and adherence to recommendations.

Relational communication for behaviour change

Whenever we talk about helping someone to change, we are probably fulfilling all of the following roles at different times:

- Consultant – someone who's primarily involved in giving advice. Behaviour consultations involve giving considerable advice in the form of recommendations.
- Counsellor – someone involved in psychological/emotional change. In human counselling it very often emphasises listening and facilitating rather than giving advice.
- Coach – someone involved in skill development. The development of new skills is certainly something behaviourists teach to a lot of their clients.

However we describe ourselves and whatever our consulting model, using client-centred counselling skills as an approach to communication fits across all of these roles and can be very useful in helping us to achieve the best outcomes for our human and horse clients. Moreover, a relational approach emphasises the role that the interactions between two individuals plays in shaping the phenomenological experience of any helping, consulting or advisory interaction.

Instead of relying solely on directive input from the practitioner, more importance is placed on co-creation as the primary means of change, whilst still being able to make use of the specialist influence of the professional. When we're talking about supporting our clients in changing their horse's behaviour, we are always talking about the humans also changing their behaviour in some way. This in turn can challenge beliefs, values and actions which have previously, in our client's perception, succeeded in achieving their various life goals. So it's the quality of the relationship between the individuals involved that matters; this is what will affect the change we are looking for.

Transactional Analysis (TA) is a psychological model of communication based on how people relate to each other. The quality of the relationship, or what is termed transactions, between the individuals involved is what matters, and this is what has the effect of facilitating change within the person. Behaviour consultations typically involve combinations of one-to-one interactions between client and practitioner and sometimes multidisciplinary relationships between the client, the vet and ourselves, or if not a vet then perhaps a physiotherapist or groom/yard manager. Our treatment plans recommend changes to the horse's management, their environment and their training activities in order to effect the required changes in behaviour and this, more often than not, requires a change in human behaviour. If we have an understanding of the psychological concepts of Transactional Analysis and their effects on human behaviour, then we can better understand and manage both the explicit and the underlying aspects of interpersonal communication. This understanding allows us to make a clearer transfer of our message to the other person. It helps us to nudge along the changes in human behaviour, environment, daily management, training activities or the relationship between horse and owner that we want to happen in order for the horse's life to improve.

The Transactional Analysis model provides an effective vehicle for ensuring authentic communication, underpinned as it is by the core beliefs that:

- people are ok
- people can think
- everyone is capable of change

We all have emotions and they accompany us when we talk to people in any social or professional situation. It is helpful to understand what may be going on for ourselves and others "below the waterline" – the things that people don't tell us but that may be affecting their actions or their ability to change. The TA model thinks of communication between individuals as ways in which we perceive the worth of ourselves and others (whether we, or others, are "ok" or "not ok"), and it also recognises that we all have three internal ego states that we hold psychologically within ourselves. These are described as our Parent, Adult and Child ego states. Transactions (communication) between the Adult ego states of two people are based on two rational "here and now" views of the world and they are seen as the ultimate aim in communication, problem-solving and decision-making. The Parent ego state is divided into two sections, the first is the Critical or controlling Parent, bound by rules and full of the "should" and "ought" to beliefs that we learned as children and we continue to carry around with us in adulthood. The second part of the Parent ego state is the Nurturing Parent, concerned with keeping the peace as well as offering a safe haven and unconditional love. The Child ego state develops from early childhood experiences and is part of an earlier self but one that still shows up regularly in our adult lives as irrationally fearful or rebellious, flying into a tantrum, expressing free and spontaneous feelings, natural curiousity or acting without considering consequences. It is the Parent and Child ego states that can intervene and create misunderstanding because these ego states consist of unhelpful messages or things we learned from our parents and other authority figures that were useful to us when we were five, but aren't much help to us when we're 35. When our inner Parent or inner Child tries to get in on the act of our adult communication this often results in misunderstanding, miscommunication or inappropriate emotional responses. Our ego states and those of the individuals we interact with will influence our interpersonal communication, so it's helpful to know about them conceptually and the effect they may have on the way we communicate with others and how others respond to us. Being aware of which ego state we are in and which one the other person is speaking to us from is very useful indeed. The most helpful ego state to be in is our Adult if we are going to be at our best in helping people to help their horses.

Now let's move on to another key concept in Transactional Analysis theory, which is the recognition of four different life positions (Table 11.1); we adopt these early on (but they're also fluid throughout life) and they have implications for how we operate in our world and how we communicate with others. The four life positions in TA are:

I'm ok, you're ok (healthy)
I'm ok, you're not ok (one up position)
I'm not ok, you're ok (one down position)
I'm not ok, you're not ok (hopeless or despairing)

We relate to people from one of these four positions. Only the I'm ok, you're ok life position is healthy and functional. We can get an idea about what life position we may be at through our emotional responses to our interactions with others. Crucially in terms of our own life positions, they reflect how we may be feeling ok or not ok about ourselves. If we are feeling "not ok" about ourselves and/or the other person (the horse's owner, for example) we may experience shame or think that we're a bad trainer or that we're giving poor behaviour advice, or we find ourselves blaming the owner for something – in any of these cases we won't be at our most effective in terms of communicating and influencing.

Let's take a look at transactions in communication. We know that effective communication is an essential skill involved in helping people to help their horses. In Transactional Analysis, communication is made up of complementary transactions Adult to Adult, or Parent to Child (and vice versa), which are all emotionally satisfying and congruent. But it is only Adult to Adult transactions that hold the essential authenticity that enables us to communicate clearly with others based on an objective appraisal of reality, and it is these transactions that we are

Table 11.1 Life Positions in Transactional Analysis

	You're ok	You're not ok
I'm ok	**Confident** I'm confident in my knowledge and abilities and I respect and acknowledge the other person's competence, with no sense of competition.	**Competitive** I know more than you, I can do my job better than you, you don't know what you're doing, you're incompetent, you don't know enough to be as good as me.
I'm not ok	**Inferior** I'm not confident, I don't know enough, I can't do my job, I can't do what's expected of me, I can't do this as well as you, I feel inferior to you.	**Hopeless** I don't think either of us knows enough to be doing this properly, we will fail at this task, neither of us has the skills or knowledge to succeed.

aiming for in our interactions with others. A simple example of a complementary transaction from Adult to Adult would be when one person asks "how are you?" and the other person responds, "I'm fine thank you". An example in a training or behaviour situation might be when you ask your client, "how are you getting on with the exercises I gave you?" and the client replies, "quite well overall. I enjoyed doing the target training but I found the timing quite difficult at first, but I think I'm getting better at it". All very realistic, with mutual respect.

On the other hand, we also have the less helpful crossed transactions, for example from a Parent to a Child ego state by one person but responded to as Adult to Adult by the other person. A basic example of a crossed transaction is where one person asks, Adult to Adult, "how are you?" and the other person has an emotional response as if they are a sulky child reacting to a parent and they say "like you care". A practical example of that in a behaviourist's work might be that the practitioner comes along to their client and from their Adult ego state they ask the client how they got on with the behaviour modification plan, but the owner of the horse maybe interprets that as a threatening challenge to their ability, or they feel guilty because they haven't carried out the BMP. In this instance they've connected with those old parental messages that are about judging and feeling judged. So they come up with some kind of defensive or sarcastic response that doesn't help the practitioner to feel good and doesn't help the communication to move along in a positive way, all because they have been unable to stay in their Adult ego state.

As behaviour consultants it's easy to fall into the finger wagging Parent ego state, and when that happens it should come as no surprise that your client goes into their Child ego state. This can look like agreement where they say things like "Yes, I understand all that" or "Yes, I'll do that" when actually they may not understand at all and they are just saying that to please (appease) you. Alternatively they may go into rebellious child mode and decide they don't like what you're saying and simply won't do it. Obviously, this is not the most helpful type of transaction to be engaged in and it's not going to give the best outcome in terms of compliance with your recommended treatment plan.

The drama triangle (Karpman, 1968) is a construct used to describe conflicted relationship transactions and is comprised of three roles – the victim, the rescuer and the persecutor (Figure 11.1). The drama triangle prevents healthy communication and inhibits problem-solving. The Victim role is characterised by an attitude of "poor me", the Rescuer role is, "let me help you" and the Persecutor role is, "it's your fault" or "you've got it wrong". As individuals we are all likely to have a tendency towards playing one of these roles and perhaps we could think that anybody who has chosen a clinical behaviour career pathway is likely to have a tendency towards the role of rescuer, not least because many enter the profession thinking "I've got all this knowledge about how to help horses and make their lives better and I want to share it". The enthusiastic behaviour consultant turns up on the yard armed with all their knowledge, experience and information which they are happy for the owners to make use of and then they get surprised when people say "well I'd rather ask my instructor" or "I'd rather ask my friend in the next stable what they do" or "I'd rather ask the yard manager". At this point it can be very easy to fall into the role of the victim and feel that the owner is persecuting you! – "they're not paying attention to me", "they're dismissing/discounting me when I've done all this training and got all these qualifications".

As behaviourists, we ourselves need to be mindful that we are not positioning other professionals as the persecutor. If we feel confused, frustrated or uncomfortable about a case it can be helpful to reflect on whether we have adopted one of the game-driven roles on the drama triangle and how this may be hindering our efforts to help the horse and their owner. When we are collaborating with other people, it's important to notice how we are feeling. Are we feeling "not ok" about ourselves or the other person? Do we feel as if we're in our helpful Adult ego state? Are we experiencing crossed transactions? It is then helpful to consider empathically what might be happening for the other person and how they too are feeling about the interaction.

In contrast to the drama triangle, the winner's triangle (Choy, 1990) empowers communication and results in effective problem-solving. Holders of the three drama triangle roles are encouraged to recognise, accept and adopt the positive (often hidden) aspects of the role they inhabit – vulnerability (victim), assertiveness (persecutor) and caring (rescuer). When our professional interactions don't seem authentic, or when we find ourselves confused or frustrated by conversations with colleagues or clients, it is likely that, probably outside our and the other person's awareness, we have entered onto one of the points of the drama triangle. The first step in moving from the drama triangle to the winner's triangle and experiencing a more effective dialogue is to notice that this is happening. That's all, just notice. We can become

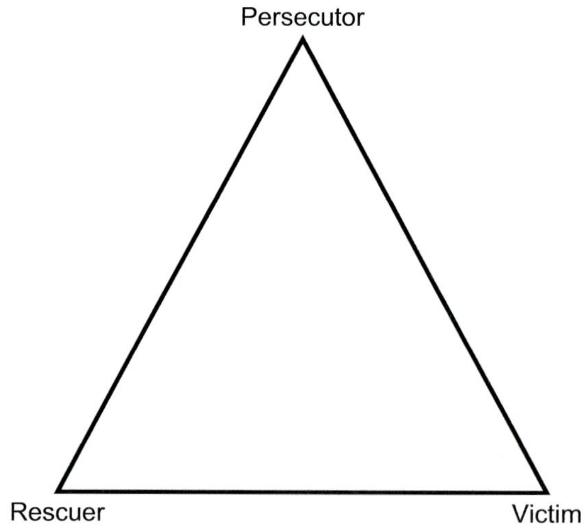

Figure 11.1 The Karpman drama triangle.

aware of when we are in our Adult ego state, and, if we notice we're not in that mental place, we can consciously move to that role. Simply becoming aware of and understanding these drama patterns is a good first step for all of us. We can then choose to change (break free from these patterns) and to step off the drama triangle. The key is to make a change from external referencing (which is where we look outside ourselves for approval, answers, solutions, love, identity and happiness) and to start internal referencing by developing our own internal agency and self-efficacy. From this position we are then better able to recognise that healthy relationships, whether personal or professional, include some aspect of authenticity, respect, sharing and open communication. To move towards this we must be willing to stay fully engaged, present and congruent.

By noticing and asking ourselves questions about how we are experiencing the interactions we find ourselves in – what position we may have stepped into on the drama triangle or how we might be participating in crossed transactions to create a difficult situation – we can explore what we really want from ourselves and others and we can take action within ourselves to step into a better functioning role, which always involves being aware of our Adult ego state and operating from that psychological place, giving us the best chance to communicate effectively with our client and to support them in supporting their horse.

Human behaviour change

A relational or collaborative approach puts the horse's caregiver at the heart of owning the resolution to their problem, but at the same time it also allows helpful input from specialist service providers such as vets, physiotherapists or other paraprofessionals. In practice, if we dive in with a plan before our client even understands or is truly aware that something needs to change they are unlikely to fully engage or comply with any change programme. If you think about things you have tried to change in your life, was it easy to make the change? Did you have a reason for wanting to make the change? Did you have the knowledge to make the change? Changing our own habits or beliefs can be hard, and yet so often we expect people to change their behaviour towards their animals simply because we know something they don't. We think we just need to give them the information and they will see the light. For example, we might want a livery yard to move from housing their horses singly to keeping them as a group. We tell them about the welfare benefits of group housing, we explain the scientific research and we describe how to make the change from individual stabling to a barn system. We give people the information but they still don't make the change. It's not always easy for people to change their behaviour for the benefit of their animal if it goes against their beliefs and how they've always done things. We are used to thinking about training our horses to perform in our chosen discipline or retraining a behaviour we don't want, but what about re-educating the human trainers? Despite many owners talking eloquently about their passion for the horse there is a gap between the advancing scientific knowledge around horse wellbeing and the practical application of these findings in everyday life. And the main reason we need to understand human behaviour and human behaviour change is that the root cause of most welfare/wellbeing issues is people doing, or failing to carry out, certain care, management or training practices.

We can consider human behaviour change in four broad areas:

1. The process of change
2. The psychology of change
3. The environment for change
4. The ownership of change

The process of change

The process of change recognises that behaviour change takes place in stages, reflected in Prochaska and DiClemente's (2005) transtheoretical model of behaviour change (Figure 11.2). The first stage is where the person is aware of conversations among friends or in social media, such as discussions about moving to group housing of horses rather than solitary stables, but hasn't really considered it for their horse. The second stage is where the person starts to contemplate the consequences for themselves and their horse if they moved away from individual stables to group housing. By the third stage the person is more actively looking at options/opportunities for group housing in their local area or at their current yard. Stage four is the action stage. The person stops stabling their horses individually and instead moves them to a group housing setup. The final stage is the maintenance stage where the person's beliefs and actions have changed and they now automatically choose group housing over individual stabling or turnout. In practice, if we go in with a treatment plan (highlighting group housing) before our client even understands or is truly aware that something needs to change, they are unlikely to be compliant in any change programme.

As part of the process of change we also need to address the capability, opportunity and motivation required to effect behaviour change, often referred to as COM-B (Michie *et al.*, 2014). It's an approach that can be used to close the gap between research evidence and the beliefs held by people who care for horses. Capability refers to the knowledge and physical skills to undertake what is outlined in the treatment plan. Does the caregiver understand or

Figure 11.2 The transtheoretical model of behaviour change. Diagram by dizain.

have the ability to train a horse to load into a trailer, for example? When looking at opportunity we need to consider things like finances, social norms and environment. For instance, it is still considered normal for horses to be housed in stables and turned out in individual paddocks, and many livery yards do not offer any alternatives. Will there be social support or social pressure that will impact on compliance? In terms of motivation, most human behaviours are habits and so strategies that fit in with a client's normal routine will be easier to implement. However, helping clients to reflect on what they want/no longer want from their horse can also harness the motivation for change. Explaining that if you provide group housing for your horse they may be easier to handle/train and therefore more likely to be successful in competition can provide the motivation necessary to make and sustain the change in some clients. Another person might be motivated by a belief that they are making their horse happy by bringing them in to a nice, warm, cosy stable – how do we educate them as to what "happy" looks like for their horse? While capability, opportunity and motivation all influence behaviour, the levels of capability and opportunity are considered to be the drivers of motivation, with motivation ultimately determining whether a behaviour is executed. Simply put, the more capable the individuals are, and the more resources and support they have at their disposal, the more motivated they will be to execute change.

There is no point expecting a change programme to be successful if there is no capability, opportunity or motivation, and we could be wasting a lot of time and energy trying to bring about changes in human behaviour that were never going to work if these three conditions are not present. If you've identified a capability issue you need to look at training or education. If the barrier is around opportunity then you might be looking at changing the environment, or leveraging social influence through modelling or demonstration. Engaging owners in the treatment plan whilst understanding their capability, opportunity and motivation to change their behaviour regarding their horse's care will increase compliance (Daniels *et al.*, 2023). We need to establish what might stand in the way of changing equestrian practices. Is it a lack of knowledge or skill? Does the physical environment or social context prohibit the execution of the behaviour? Or do ingrained habits, instincts or even a lack of planning prevent meaningful change? A thorough analysis aimed at determining any potential barriers to the desired target behaviours is an essential first step towards mapping out an effective behaviour change strategy.

The psychology of change

Understanding psychology is another key factor in driving change. We need to look at what is happening psychologically when we try to persuade people to change their perspective or the way they do things. Four conclusions have been found to hold true across a wide range of social psychological studies. Firstly, confrontation is the biggest predictor of failure to change. Secondly, an empathetic approach is the biggest predictor of success in bringing about behaviour change. If you fundamentally believe that people are ok and that everyone can change, you have the best chance of succeeding. Next, the phenomenon of psychological reactance means that if someone gives you reasons why you should change, you are likely to highlight the reasons why you shouldn't, and in doing so you rehearse these beliefs and make them stronger. Finally, we need to be aware of what's called the righting reflex (think a cat landing on its feet after a fall): we all automatically like to offer *our* solutions, but then those solutions are not owned by the person we are trying to influence. We tend to think that people are like us, so we think that they will be motivated by the same factors that motivate us, but we are all different. Consider that we might stop offering solutions so readily. Different people need different types of information to change their behaviour. It's no good just shouting one message to everybody.

We cannot assume that we know why people are doing or not doing something. Get that wrong and your whole strategy is flawed. We need to understand the barriers that stop people making changes; understand their beliefs and the "lens" that they see life through. Our individual experiences and our relationships mould the way we look at life and the reasons for our actions. Understanding these aspects of human psychology helps us to better communicate and to drive change.

One last point on the psychology of change is to mention confirmation bias; this is a psychological term to describe that we are more likely to take on information if it aligns with what we already believe and know. So if what we need to say to somebody doesn't align with their knowledge and beliefs, we need to be aware of this and work out how we can be "heard". It's worth noting that our decisions are not as autonomous as we would like to think.

The environment for change

Consider creating an environment for change by shaping the path. Environment has a huge impact on someone's ability to change: home life, relationships, employment status, life down on the livery yard or in the riding club. And the cultural and legislative environment affects the likelihood of successful change too. We can shape the path to change by giving people information, a clear goal and a destination but we also need to get their emotional engagement, through building rapport and making our explanations resonate at an emotional level – how will it feel to reach the goal? Tools we can use here include prompting, identifying enabling factors, follow up, support networks and family networks, and helping people to form habits that will support long-term change. We also need to understand barriers to change. Why don't horse owners follow vets' or behaviourists' recommendations? Why don't owners always follow a training plan? Why do people often carry on the hard way? It's better to use a proven method to truly analyse these behaviours rather than making assumptions, because wrong assumptions lead to ineffective communication. For example, if we assume that people are overfacing their horses during misunderstood attempts at "desensitisation" because they don't know about the phenomenon of psychological flooding, and you talk to them about "the problem of flooding", then that might not remove any of the barriers to change, if that level of knowledge wasn't the barrier. Maybe they do know about flooding but hold other beliefs such as "it's important for me to have a quiet, obedient horse", so it would have been more helpful for you to pitch your conversation at that level instead.

Ownership of change

The last principle of human behaviour change is ownership of change. There is a saying, "Tell me and I forget, show me and I remember, involve me and I truly understand", which perfectly illustrates this principle of change. We need to truly appreciate the relevance of the desired behaviour change to ourselves in order for change to happen. If we understand that people don't always learn and change if they are told what to do through a typical top-down educational outreach type of approach, or if they are only shown what to do through demonstration, but are more likely to change if they are truly involved in the process of change, then we can facilitate that change. This process involves enabling people to explore issues and come up with solutions themselves rather than "training them" to implement a preconceived solution.

We need to apply the principles of human behaviour change at all stages of our interactions, starting from the initial phone call or email if that's the first point of contact with our client. We can train ourselves to empathise emotionally in our first comments, so for example we can say, "that must be so frightening for you, don't worry, we can work to find a plan that fits with you". By acknowledging the emotional element of the individual's problem, and at the same time setting clear boundaries by explaining how we work and our clinical practice framework, we are managing expectations in a fair and reasonable way. Whether we're working with one-to-one clients, groups or organisations, we need to stay flexible. Find the balance between researching for the situation you think you're going to find yourself in, but don't assume anything, and don't be so fixed on a "script" that there's no room left for dealing with the actuality of what you find in front of you.

Clinical supervision

At this point I want to emphasise the need for supervision. As well as being a clinical equine behaviourist I am also a practising psychotherapist and as such I am required to have clinical supervision with a set number of hours as a ratio of my practice hours (BACP, 2020). In fact, it is a mandatory requirement for all human counsellors and therapists who work under a professional code of conduct to undertake supervision. I've observed time and again that we, as clinical behaviourists, are dealing with emotional material from our clients and their horses but have little professional opportunity to process this. A partner or good friend to talk to is great in theory, but not sufficient for the role we are undertaking as behaviour therapists, and we are bound by client confidentiality. When we talk to a client we are in effect holding space for them and taking in everything they are telling us, including the emotional content – emotional pain, distress, anger, fear, confusion, frustration. Add to this the emotional content we also take in from the horse, as well as the tendency for many of us to take on a "rescuer" role, and the toll can be heavy. Many behaviourists come into the profession because they don't want to see horses in pain or distress, that's their motivation, and yet what happens is that in reality, every day they see horses experiencing pain/distress/fear/anxiety. In order to be able to maintain resilience over the long term, a therapist needs a place to process and discharge what is taken in from clients, both horse and human. From my perspective, this lack of mandatory supervision contributes to levels of burnout among behaviourists. Some who have worked hard for three years for their undergraduate degree, then a further two years at postgraduate, may actually only practice for 18 months before leaving the industry. So my take-home message for all clinical behaviourists, whether newly qualified or seasoned practitioners, is to put in place regular, high-quality, professional supervision. Not only does clinical supervision provide a safe space to discuss and reflect on your clinical practice, your work with clients and your role as a professional practitioner, it's also a place to explore any challenges you are facing and continue your development. For provisional or candidate behaviourists it's an opportunity to become more familiar with clinical behaviour working practices and systems and helps to develop confidence in your professional skills. The British Association for Counselling and Psychotherapy (BACP, 2021) has a lot of resources available to the public as well as to members if you want to find out more about clinical supervision.

References

BACP. (2020). *Introduction to Supervision (Public Version)*. Available at www.bacp.co.uk/media/10210/bacp-intro-to-supervision-caq-gpia064-nov20.pdf

BACP. (2021). *Peer Supervision within the Counselling Professions*. Available at www.bacp.co.uk/events-and-resources/ethics-and-standards/good-practice-in-action/publications/gpia121-peer-supervision-within-the-counselling-professions-fs/

Choy, A. (1990). The winner's triangle. *Transactional Analysis Journal*, *20*(1), 40–46.

Cooper, M. (2024). *The Handbook of Person-Centred Psychotherapy and Counselling*. Bloomsbury Publishing.

Daniels, J. T., Busby, D., Chase-Topping, M., & Brown, S. M. (2023). I wish he'd listen: Client-centered interviewing approaches are associated with higher compliance with behavioral modification advice in pet dog owners. *Journal of Veterinary Behavior*, *63*, 22–30.

Egan, G. (1990). *The Skilled Helper: A Systematic Approach to Effective Helping*. Thomson Brooks/Cole Publishing Co.

Human Behaviour Change for Animals (HBCA). (2019). Available at www.hbcforanimals.com/

Karpman, S. (1968). Fairy tales and script drama analysis. *Transactional Analysis Bulletin*, *7*(26), 39–43.

Michie, S., Atkins, L., & West, R. (2014). The behaviour change wheel. In *A Guide to Designing Interventions* (1st ed., pp. 1003, 1010). Great Britain: Silverback Publishing.

Prochaska, J. O., & DiClemente, C. C. (2005). The transtheoretical approach. *Handbook of Psychotherapy Integration*, *2*, 147–171.

Stewart, I., & Joines, V. (1987). *TA Today: A New Introduction to Transactional Analysis*. Nottingham: Lifespace Pub.

Wolframm, I. A., Douglas, J., & Pearson, G. (2023). Changing hearts and minds in the equestrian world one behaviour at a time. *Animals*, *13*(4), 748.

Part 5

Psychopharmacology awareness

Chapter 12

A behaviourist's guide to psychopharmacology

What is psychopharmacology?

Psychopharmacology is the study of the use of drugs that affect an individual's mind and their behaviour. These drugs affect the brain's physiology (normal functioning) and they also interact with the endocrine system (glands and hormones), resulting in changes in behaviour and motivation. They can be a useful addition to the treatment of behaviour problems in animals but it is uncommon for psychoactive medication alone to provide a resolution. In the majority of cases, treatment is most effective if medication is used in combination with environmental management and a behaviour modification programme. It's also worth noting that much of the data on the efficacy and safety of psychoactive drugs is derived from human psychiatric use and then extrapolated for use in veterinary clinical behavioural medicine. (Crowell-Davis *et al.*, 2019).

What is a psychopharmaceutical?

A psychopharmaceutical is a drug that has an effect on an individual's mental state. You may also hear them referred to as behaviour modifying pharmacologic agents, or psychoactive/psychotropic drugs. The reason we need different types of medications is that whilst fear, anxiety, phobias and repetitive disorders are probably related to each other at the neurochemical level, they may have different underlying mechanisms, hence the need for different drugs to treat them.

It's important to appreciate that any psychopharmaceutical may be considered a chemical restraint, particularly if used not to treat an underlying condition but to suppress a horse's unwanted behaviour or reaction to their environment or what is happening to them (e.g. a veterinary procedure). This may be warranted in emergency, time-critical or dangerous situations as a form of urgent behavioural first aid, but otherwise needs very careful consideration. They are not a replacement for a behaviour modification programme or to compensate for the lack of species-appropriate opportunities for behavioural expression consistent with the five domains of animal welfare (Mellor *et al.*, 2020). Interestingly, scientific literature demonstrates that most behaviour problems, even when referred by vets for behaviour-only therapy, are often caused by or at least partially influenced by physical problems (Jonckheer-Sheehy, 2024a).

DOI: 10.1201/9781003581390-17

What is my role as a behaviourist in the use of a psychopharmaceutical?

Prescribing or recommending any sort of medication is the responsibility of the horse's referring vet and outside the scope of practice for non-veterinary behaviourists. However, the UK Animal Behaviour and Training Council (ABTC) and Fellowship of Animal Behaviour Clinicians (FABC) codes of conduct (ABTC, 2021; FABC 2024) do allow clinical behaviourists to discuss the general use of psychoactive medications as part of a behaviour modification programme but without going into any detail about specific drugs. It's really important to note that the terminology used is "discuss", not "advise", and refers to both conversations with clients and those with the referring vet, who holds the responsibility about whether to and what to prescribe.

Psychopharmacology and other adjunctive medical treatment may be discussed with the referring vet but only with reference to the evidence base; that is, current and established scientific literature. A non-vet clinical behaviourist cannot advise the client or the vet on the specific appropriate treatment or a specific class, type or dose of medication as this remains the full legal responsibility of the referring veterinary surgeon under the Veterinary Surgeons Act 1966. In addition, you cannot claim competence in the clinical use of psychopharmacology or other adjunctive medical treatment (even if you have first-hand experience) as a non-vet clinical behaviourist.

When might the use of a psychopharmaceutical be indicated?

In reality, psychopharmaceuticals are not widely used for behaviour problems in horses. This may in part be due to the lack of scientific literature or strong evidence to support their use in horses from both an efficacy and a safety point of view. Where they are used, they are generally aimed at the short-term management of behaviour cases where the emotional state of the horse implies that their suffering is extreme or that their behaviour is dangerous to themselves or others; that is, when it's necessary to reduce reactivity, fear, anxiety or aggression in an emergency situation. What this looks like in practice is an animal who simply cannot adapt to their environment in the immediate moment (Jonckheer-Sheehy, 2024b).

Another consideration to be aware of, potentially to discuss but not advise on as a non-vet behaviourist, is competition rules and regulations. The International Equestrian Federation (FEI) has a published list of prohibited substances. Medications on the controlled list are deemed to have therapeutic value or are commonly used in equine medicine and may have the potential to affect performance. These medications are prohibited from competition and a withdrawal time must be observed. Beyond this, banned substances are those not permitted for use at any time. They are deemed to have no legitimate use in the competition horse and/or to have a high potential for abuse. The British Horseracing Authority (BHA) also has a list of controlled medications and their detection times but at the time of writing no list of prohibited substances.

In my experience as a clinical behaviourist, unless there is a clearly evident immediate welfare or safety reason for considering behavioural medication, the typical time to consider if the addition of a psychopharmaceutical is indicated is after the initial stages of the behaviour modification plan have been completed (attending to friends, forage and freedom, adapting the environment, downtime and enrichment activities). After these stages of the treatment plan, and once you have started the mental and emotional rehabilitation exercises using desensitisation and counterconditioning, if you are still not seeing any improvement and attempts to improve the horse's emotional and mental state are showing no effect, then it might be time to have a discussion with both client and vet about the possible use of a psychoactive agent. This is with the proviso, however, that every clinical behaviour case must be assessed and evaluated on its own merits including, essentially, investigation of any pain-related factors (and their treatment, if identified). In my experience, it is unlikely for there to be no discernible improvement if the client has followed the initial steps of their behavioural treatment plan; however, this does not discount concurrent first thoughts about psychopharmaceutical assistance in cases where the horse is so stressed or anxious that they are unable to engage with the behavioural treatment plan in any effective way. These considerations may also relate to cases where the horse's environment or management or the client's ability or motivation are not sufficient, in terms of implementation of recommendations, to achieve a minimum level of effectiveness. In such cases behavioural medication may help by bringing arousal down to a level where some "brain bandwidth" is made available for learning and behaviour modification to be able to take place.

Another crucial point to note here is the interaction between health and behaviour, and the importance of assessing (or reassessing) for pain, even at the same time that you may be considering behavioural medication.

There is also an overarching ethical question about the use of psychopharmaceuticals at all, which applies to other companion species as well as horses. In a situation where one species lives their lives bounded by the social mores, resources, goals and mindsets of another (humans), at what stage does one intervene medically to do something that is going to affect the functioning of that animal's brain? Behaviourists who work with dogs have been aware for some time that for serious behaviour issues, the underlying fear, panic, anxiety or phobia is affected by physical changes that can have an inhibitory effect on learning, and if these emotional states cannot be managed to a level where learning can take place such that it enables a domestic animal to thrive in their environment, it may be appropriate to consider psychopharmacologic agents to facilitate this process. At the same time, we can usefully bear in mind the role of policy-makers, legislators and social philosophers in the wider debate around how we expect another species to live and to perform in order to achieve our human aspirations, crystallised currently in the equestrian world as the social licence to operate and good life debates.

Your awareness of psychopharmacological medications

There are two main types of psychopharmacological medications used in horses that you are likely to come across when working in the behaviour field:

- sedatives and tranquilisers
- antidepressants (TCAs and SSRIs)

Sedatives and tranquilisers

Sedative drugs induce sedation, which is described as a state of relaxation or sleep. In contrast, tranquilisers are generally thought to reduce anxiety without drowsiness or sleepiness. In that sense sedatives are likely to have a more profound effect than a tranquiliser. The most commonly used tranquiliser to treat behaviour problems and perhaps the most recognisable is acepromazine (ACP or ACE), one of the phenothiazine class of drugs. In humans it is classed as an anti-psychotic and in horses is used to calm nervousness and reduce aggression, also used in horses on box rest. The benefits of ACP are that it is short acting with mild to moderate sedation at best, although the sedative effects do vary according to the individual horse. It is also a mild muscle relaxant but has no analgesic (pain relieving) or anxiolytic (anxiety reducing) properties. What this means is that even if a horse is on ACP they will still react emotionally if they are in pain or are fearful. The main brands of ACP preparations that you are most likely to come across include Sedalin, Equipazolone paste and Tranquigel.

Interestingly the European Society of Veterinary Clinical Ethologists have contraindicated ACP in small animal medicine where anxiety is present due to concern about its effect as a "chemical straitjacket". In other words, the animal is fully aware of everything going on around them but they are prevented from reacting to it (Pearson, 2024).

In terms of sedation, alpha 2-adrenoceptor agonists xylazine, detomidine and romifidine are classified as analgesics (pain relieving) as well as sedatives and were originally developed for use in humans with hypertension. They are all given as an injection so are more likely to be used in emergency situations rather than in medium- to long-term behaviour modification. Detomidine hydrochloride (Dormosedan) can also be administered as an oral gel via the cheek mucosa, which is helpful if you have a horse who is particularly fearful of injections. A potential disadvantage of these types of drugs is that excessive kicking reflexes have been associated with their use.

In the opioid class of drugs, butorphanol (branded as Torbugesic) has both sedative and analgesic properties but without the loss of proprioception (awareness of the position and movement of the body) or consciousness. It can cause excitement/increased locomotion if given alone and so is more commonly used in conjunction with one of the other sedatives discussed previously.

Antidepressants

There are two main types of antidepressants used in horses – tricyclic antidepressants (TCAs) and selective serotonin reuptake inhibitors (SSRIs). TCAs (clomipramine, amitriptyline) were originally used as anti-psychotics in humans and were once typically used in horses to decrease anxiety, hyperreactivity and aggression. They have since been superseded by the SSRIs in both human and veterinary medicine, although they have not been thoroughly evaluated in horses as yet. Fluoxetine (Prozac) and paroxetine (Seroxat) are the most familiar form of SSRIs and are likely to become more widely used given their prominence in both human and small animal medicine (McGreevy, 2012), although no veterinary form is currently licensed for horses in the UK. In terms of research, a retrospective evaluation of fluoxetine use in 95 horses found that it was a useful tool to treat equine behaviour-related problems when prescribed to facilitate box rest and the resulting behavioural problems but that further controlled behavioural studies are needed (Fontenot *et al.*, 2021).

Trazodone

At this point it's worth mentioning trazodone, a drug which is becoming increasingly popular in small animal veterinary behavioural medicine. Trazodone is a serotonin receptor antagonist and reuptake inhibitor (SARI) that is used extensively as an anxiolytic (anxiety reducing) medication in human and small animal veterinary medicine, and it's important to know that it shares many of the properties of SSRIs in the context of serotonin bioavailability. Currently there is only one published report on its use in adult horses (Knych *et al.*, 2017) and no published reports of the use of the drug in clinical cases. In fact, the 2017 study found that the administration of trazodone to horses elicited a wide range of effects and concluded that additional study is warranted before its clinical use can be recommended. However, a 2018 study looked at the medical records of horses treated with trazodone for various abnormal behaviours (box walking, circling, pacing, weaving, kicking stall, rearing and violent behaviour) whilst being box rested for veterinary procedures. In 17 out of the 18 cases trazodone, when subjectively reviewed, did cause some degree of sedation as well as modifying the adverse behaviours associated with box rest. This group of researchers concluded that trazodone may be used as an alternative oral sedative for horses who demonstrate behavioural issues while being box rested for various medical conditions (Davis *et al.*, 2018).

In reality, at the time of writing, there is very little evidence for any of the psychoactive medications being used in horses, although some studies are currently being undertaken. This means that the data is simply not available to say definitively under what circumstances they should be used, what dose is the most effective and what constitutes a safe/unsafe dosage level.

Non-prescription calming products

It's important to be clear about the difference between prescription products and non-prescription products. Prescription products are those medications you can only obtain with a prescription from a vet who has seen the horse. In contrast, non-prescription products are readily available over the counter for anyone to buy. In the same way that clinical behaviourists are not allowed to give advice on the use of prescription psychoactive drugs, the ABTC code of conduct also stipulates that CABs must not give any specific form of physical health-related advice, and this includes advice on diet, pheromonotherapy and complementary products with active chemical ingredients, without first consulting the referring vet.

Despite this, when working as a behaviourist you are more than likely to come across the use of a range of non-prescription calming products in your day to day practice. Research has shown that calming products are commonly used for behaviour modification by owners and handlers because they claim to offer an effective and non-invasive way to manage unwanted behaviours. In many cases, owners may be more willing to try behavioural supplements rather than to alter diet or training/management practices. What is especially concerning is that many of the benefits claimed by these products remain unsubstantiated and the equine supplement market is largely unregulated (Ross and Roberts, 2018).

Although many of these non-prescription calming products may have anecdotal evidence, in other words people's testimony about how well they've worked, there is actually very little scientific testing of their efficacy or safety or even if they will work at the doses being recommended on the packaging.

The most common non-prescription calming products where there is some data include:

- tryptophan
- magnesium
- alpha-casozepine (Zylkene)
- equine appeasing pheromone (Confidence EQ)

Tryptophan

Tryptophan is an essential amino acid. Essential amino acids are organic compounds that the body needs to function but that cannot be produced by the body and therefore must be taken in through the diet. Tryptophan is suggested to work as a calming agent, and there is some research in humans to support this as well as in dogs where it has been shown to reduce aggression. However, whilst tryptophan is marketed worldwide as a calming agent for fearful or excitable horses, evidence to support its efficacy or safety is inconsistent, and the studies that were done concluded that training methods/behaviour modification are, at present, more reliable ways of reducing reactivity (Grimmett and Sillence, 2005; Malmkvist and Christensen, 2007). A later study also concluded that there was little evidence for the behavioural or physiological calmative effects of supplemental tryptophan in horses (Davis *et al.*, 2017).

Magnesium

Although there is some evidence in other species, including humans, that magnesium supplementation has an effect on mood, aggression and anxiety-related behaviours, there is currently no evidence that inorganic (magnesium oxide) or organic (chelated magnesium) forms of magnesium have any effect on horse behaviour. There are many anecdotal reports from trainers and owners of behavioural improvements from magnesium supplementation but as yet no scientifically robust study to back this up. Interestingly a small pilot study looking at the effect a magnesium supplement had on reactivity concluded that although their results were inconclusive, they did suggest anecdotal claims of calming effects may warrant further investigation (Nielsen and O'Connor-Robison, 2014). In contrast, a 2018 study recommended that magnesium supplementation could not be relied on to modify reactivity in performance horses and that owners/trainers would be better advised to seek alternative strategies to manage reactive behaviour such as training and habituation to stressful situations (Dodd *et al.*, 2018).

Alpha-casozepine (branded as Zylkene)

Alpha-casozepine is a nutritional supplement which has been shown to have anxiolytic (anxiety reducing) properties. One study showed alpha-casozepine supplementation benefitted adult semi-feral ponies undergoing potentially stressful situations linked to domestication – stabling, haltering, leading, separation, health care treatments and transportation (McDonnell *et al.*, 2013). Another showed modest benefit to horses for improvement of compliance and apparent comfort with mildly aversive routine health care procedures (McDonnell *et al.*, 2014).

Equine appeasing pheromones

Pheromones are substances which are released into the environment by an animal (or human) that affect the behaviour or physiology of another. They have the potential to produce behavioural and physiological effects when perceived in the environment by other animals (typically of the same species). Pheromones can influence copulation, aggression, dam-infant bonding, familial recognition, appeasement and synchronisation of oestrus. In theory pheromones have the potential to affect the emotions of an individual in ways which may result in unconscious behavioural effects, and an effect on emotional state may result in a change in behaviour (Jonckheer-Sheehy, 2024b).

Equine appeasing pheromone (branded as Confidence EQ) is a synthetic version of a maternal pheromone released when a foal is suckling, and unlike other non-prescription calming products is applied to the nostrils (Berger *et al.*, 2013). There have been a number of studies done on the use of equine appeasing pheromones, including their effect in fear-eliciting situations (Falewee *et al.*, 2006), some of which were undertaken by the developers of Confidence EQ, and the results have been mixed. One study in 2019 looked at the use of equine maternal pheromone in foals undergoing hoof trimming and found no evidence for the effectiveness of synthetic pheromones in this situation (de Paula *et al.*, 2019).

In summary, when thinking about non-prescription calming products we simply don't have the robust validated studies to demonstrate how they work, when they work (what situations/ behaviour), what effect they have and the safe dosage rates. This doesn't mean they don't work, but until we have the research to back up the calming claims/anecdotal evidence then what the science is telling us is that the most effective course of action for the resolution of behavioural problems is to focus on the horse's health, management, environment and behaviour modification programme rather than to rely on calming products, and at the present stage of research into the use of prescription psychoactive drugs with horses, the same health warning probably also applies.

References

Animal Behaviour and Training Council (ABTC). (2021). *Code of Professional Conduct for Practitioners of Animal Training and Behaviour Therapy*. Available at https://abtc.org.uk/ (Accessed 29 April 2024).

Berger, J. M., Spier, S. J., Davies, R., Gardner, I. A., Leutenegger, C. M., & Bain, M. (2013). Behavioral and physiological responses of weaned foals treated with equine appeasing pheromone: A double-blinded, placebo-controlled, randomized trial. *Journal of Veterinary Behavior*, *8*(4), 265–277.

Crowell-Davis, S. L., Murray, T. F., & de Souza Dantas, L. M. (2019). *Veterinary Psychopharmacology*. John Wiley & Sons.

Davis, B. P., Engle, T. E., Ransom, J. I., & Grandin, T. (2017). Preliminary evaluation on the effectiveness of varying doses of supplemental tryptophan as a calmative in horses. *Applied Animal Behaviour Science*, *188*, 34–41.

Davis, J. L., Schirmer, J., & Medlin, E. (2018). Pharmacokinetics, pharmacodynamics and clinical use of trazodone and its active metabolite m-chlorophenylpiperazine in the horse. *Journal of Veterinary Pharmacology and Therapeutics*, *41*(3), 393–401.

de Paula, R. A., Aleixo, A. S. C., da Silva, L. P., Grandi, M. C., Tsunemi, M. H., Lourenço, M. L. G., & Chiacchio, S. B. (2019). A test of the effects of the equine maternal pheromone on the clinical and ethological parameters of equines undergoing hoof trimming. *Journal of Veterinary Behavior*, *31*, 28–35.

Dodd, J., Noble, G., Harris, P., Bhanugopan, M., Nielsen, S., & Doran, G. (2018). Does oral magnesium aspartate supplementation affect reaction speed in horses of different breeds? *Comparative Exercise Physiology*, *14*(Supp 1), S74.

Falewee, C., Gaultier, E., Lafont, C., Bougrat, L., & Pageat, P. (2006). Effect of a synthetic equine maternal pheromone during a controlled fear-eliciting situation. *Applied Animal Behaviour Science*, *101*(1–2), 144–153.

Fellowship of Animal Behaviour Clinicians (FABC). (2024). *Code of Conduct*. Available at https://fabclinicians.org/code-of-conduct/ (Accessed 29 April 2024).

Fontenot, R. L., Mochal-King, C. A., Sprinkle, S. B., Wills, R. W., & Calder, C. D. (2021). Retrospective evaluation of fluoxetine hydrochloride use in horses: 95 cases (2010–2019). *Journal of Equine Veterinary Science*, 97, 103340.

Grimmett, A., & Sillence, M. N. (2005). Calmatives for the excitable horse: A review of L-tryptophan. *The Veterinary Journal*, *170*(1), 24–32.

Jonckheer-Sheehy, V. (2024a). *Equine Psychopharmacology and Other Therapeutics, Part 1*. Available at www.apbc.org.uk/Events/equine-psychopharmacology-and-other-therapeutics-part-1/ (Accessed 29 April 2024).

Jonckheer-Sheehy, V. (2024b). *Equine Psychopharmacology and Other Therapeutics, Part 2*. Available at www.apbc.org.uk/Events/equine-psychopharmacology-and-other-therapeutics-part-2/ (Accessed 29 April 2024).

Knych, H. K., Mama, K. R., Steffey, E. P., Stanley, S. D., & Kass, P. H. (2017). Pharmacokinetics and selected pharmacodynamics of trazodone following intravenous and oral administration to horses undergoing fitness training. *American Journal of Veterinary Research*, *78*(10), 1182–1192.

Malmkvist, J., & Christensen, J. W. (2007). A note on the effects of a commercial tryptophan product on horse reactivity. *Applied Animal Behaviour Science*, *107*(3).

McDonnell, S. M., Miller, J., & Vaala, W. (2013). Calming benefit of short-term alpha-casozepine supplementation during acclimation to domestic environment and basic ground training of adult semi-feral ponies. *Journal of Equine Veterinary Science*, *33*(2), 101–106.

McDonnell, S. M., Miller, J., & Vaala, W. (2014). Modestly improved compliance and apparent comfort of horses with aversions to mildly aversive routine health care procedures following short-term alpha-casozepine supplementation. *Journal of Equine Veterinary Science*, *34*(8), 1016–1020.

McGreevy, P. (2012). *Equine Behavior: A Guide for Veterinarians and Equine Scientists* (2nd ed.). London: Elsevier.

Mellor, D. J., Beausoleil, N. J., Littlewood, K. E., McLean, A. N., McGreevy, P. D., Jones, B., & Wilkins, C. (2020). The 2020 five domains model: Including human–animal interactions in assessments of animal welfare. *Animals*, *10*(10), 1870.

Nielsen, B. D., & O'Connor-Robison, C. I. (2014). A pilot study to determine if a dietary mineral supplement can affect reactivity to stimuli by horses in training. *Comparative Exercise Physiology*, *10*(3).

Pearson, G. (2024). *Equine Psychopharmacology*. Available at https://event.webinarjam.com/replay/64/7vqwgumwtontnpbmq2 (Accessed 4 May 2024).

Ross, J., & Roberts, J. L. (2018). Equine calming products: A short survey into their use, effect and knowledge using a small sample of horse owners in the North of Scotland, UK. *Journal of Equine Veterinary Science*, *68*.

Appendix 1
The five domains welfare model

Adapted from Mellor *et al.* (2020)

Physical/functional domains

Domain 1
Nutrition/Hydration

Negative: Restricted food/ water intake; Poor food quality; Infrequent feeding
Positive: Sufficient food/water; Preferred food tastes; smells and textures

→ → →

Domain 2
Environment

Negative: Extreme heat or cold; Injurious hard surfaces; Restricted space; Smells of urine/excrement
Positive: Thermally comfortable; Supportive bedding; Fresh air

→ → →

Domain 3
Health/Functional Status

Negative: Acute or chronic injury; Ill health; Poor physical fitness; Lame
Positive: Injury free; Robust good health; Good fitness level

→ → →

Domain 4
Behaviour

Negative: Barren, impoverished, unvarying environment; Threatening circumstances; Isolation
Positive: Able to explore, herd and exercise extensively; Able to socialise with others and play with others and solo

→ → →

Affective experience domain

Domain 5
Mental State

Negative:
Thirst; Hunger;
Heat or cold stress;
Physical/olfactory discomfort;
Boredom, frustration;
Pain, nausea, exhaustion, inflammation, immunosuppression;
Panic, fear, anxiety, helplessness;
Loneliness, depression, neophobia

Positive:
Pleasure in eating/drinking;
Pleasure of thermal, physical, olfactory comfort;
Feeling the vitality of robust good health and physical fitness;
Rewarding engagement with exploration, herding and exercise;
Comfort and security of socialising with bonded others;
Rewards of exercising agency and having a sense of control

➡ ➡ ➡

EQUINE WELFARE STATUS

Appendix 2
Your horse's behaviour threshold

What is your horse's behaviour threshold and why does it matter?

The idea of a threshold is often used when talking about aggression, fear and reactivity. It's also important to know about thresholds when bringing on young horses, and it's useful for teaching excitable horses to be calm or for helping anxious horses to be more resilient.

Understanding about horses' thresholds helps you to be a better trainer and will give you a more relaxed and resilient horse.

When you cross a threshold, you move from one area to another. In a behaviour threshold your horse crosses from one emotional state to another, for example from calm to excited, or from relaxed to fearful. If you have a horse who worries about the scary corner of the school, for example (every arena's got one!), you have probably seen that they move from appearing to be calm and relaxed, to feeling tense; they may slow down as they try to avoid going towards that area, and they may stop listening to your leg, or they may spin and becoming difficult to ride, not responding to your cues. When this happens, you know they have gone over threshold.

Often we think of over-threshold horse behaviour that looks like rearing, napping or bolting, but there are other ways that a horse's behaviour tells us they are over threshold, for example:

- shutting down or appearing to freeze
- being overexcited (for example, doing that wonderful passage past the unexpected donkey when you haven't even taught them that in the school!)
- being so distracted that they cannot pay attention to you
- licking and chewing
- yawning
- nipping
- shaking their head
- rubbing their head on something, or rubbing their nose on their knee

You may notice that your horse stops taking food treats, stops interacting with you or suddenly seems quieter, less active, than usual. These may be signals that your horse is approaching a threshold, or has even already moved from a comfortable emotional state into an uncomfortable one.

Over-threshold behaviour is not bad behaviour

When a horse is over threshold it generally means to us that they are behaving in a way that we don't like. From their side, it can often mean that they are experiencing a state of anxiety, fear or distress.

When a horse crosses an emotional threshold, certain physiological and psychological effects begin to take place. The horse may breathe more heavily, heart rate may increase. A horse who is over threshold is *reacting* rather than thinking. They are in the fight or flight state that you have probably heard of. There are other "Fs" as well: freeze (with this, it can seem as if they are holding their breath, rather than breathing heavily), fidget (e.g. walking a full semi-circle at the end of the lead rope if tied up) or fawn (appeasing/submissive). They may appear not to listen to you, and this may be because they can't actually hear you at that time. When a horse is over threshold, they cannot learn a new or different behaviour. A horse will not be able to learn until they have come back down under threshold and are feeling relaxed and calm.

Thresholds can change

An emotional or behaviour threshold can change from minute to minute and from one situation to the next. The setting for a horse's threshold at any moment depends on a number of factors:

The number of triggers: the more triggers, the greater the risk that your horse will go over threshold. A lot of small triggers at the same time, or even just one after the other on the same day, can put a horse over threshold. Watch this video on "trigger stacking" for more information on this: https://youtu.be/BtPb_lvVkSk (copy and paste the link into your browser if necessary).

A trigger is not always something scary; it can be anything that increases your horse's arousal or excitement level. For example, galloping with another horse can cause some horses to go over threshold, something to be aware of when you're out riding, especially if you're a novice or the horse you're on is new to you.

Proximity, or closeness to the trigger. In most cases, closer is more challenging, but not always – sometimes something further away may be scary because the horse isn't sure what it is. This is often the case when a horse spots something on the horizon of a hill, for example.

Frequency (how often the trigger occurs). If a horse faces the same trigger repeatedly, especially in a short period, they may react more strongly.

The intensity of the trigger: for example, if the trigger is a sound, then the volume or duration might affect the horse's reaction.

Being hungry, thirsty, tired, or in pain can all have an effect on a horse's threshold.

Accumulated stress can also affect a horse's threshold. For example, if your horse has a fun but stressful weekend at a show or a fun ride (this may include the travelling there and back), they may go over threshold more quickly if spooked by a loud sound on Monday morning. (It may take several days for their stress hormones to return to normal, so accumulated stress responses are not always easy to trace.)

You can help your horse stay under threshold

To help your horse stay under threshold, you can learn what types of things might be triggers for your horse, and you can find your horse's threshold for different triggers. Anything that creates stress, high arousal or overexcitement is a possible trigger. Identify both positive triggers (like moving fast, or play biting) and negative stressors (like scary objects, certain situations or loud noises).

Learn to read your horse's body language and what signals show that your horse is going over threshold. For some horses, you may notice tension, stiffness, growing a hand ("big horse"); other horses become more excited, and some may try to move away or start pawing the ground. Most horses grab at hand-delivered food and take it more roughly when they are getting close to threshold.

Managing your horse's threshold

Stay focused on your horse and read their body language, look for stress signals – yawning, lip licking, tension around the mouth, triangulated upper eyelid, ears back, looking away, stiffness, slower movement. Look for changes in your horse's behaviour or energy levels.

This matters because every time your horse goes over threshold in response to a scary trigger, they are becoming more sensitised to that trigger and will eventually (or quickly, sometimes even in a single scary or distressing event) develop an automatic response so that they will then show that unwanted behavioural response every time they see the same trigger. Think of horses who always spin and run a few metres before stopping and checking out a tractor, a bus or a person with an umbrella, for example. If they always react in this way then they have developed what is called in psychological terms a conditioned emotional response, and this becomes very difficult to "undo" when you are trying to eliminate unwanted behaviour.

If you are working with a behaviourist on changing your horse's fear or reactivity response, going over threshold frequently will slow down your horse's progress.

If you notice your horse's arousal increasing, take action; don't wait for them to go over threshold. If your horse is getting close to threshold (the amber zone in the following picture), you can:

- √ Create distance between your horse and the trigger by moving further away.
- √ Use targeting exercises, if you have previously taught your horse to target.
- √ If necessary, leave the situation altogether.

Training and behaviour modification are important solutions to help your horse stay under threshold. Over time, this can change a horse's threshold levels around certain triggers. Plus, when you are around a known trigger, where over-threshold behaviour is a risk, you will know what to do to. Your behaviourist will help you with this.

If your horse goes over threshold (red zone – danger):

- √ Get your horse out of the situation immediately. This is not a time for training, learning, or fixing problems.
- √ Take note of all of the factors that led to your horse going over threshold.
- √ Make a plan for the next time you are in that situation, so that you can prevent it from happening again if at all possible.

Helping your horse to stay under threshold puts them into a mental and emotional place where they are able to learn new behaviours or different responses, which can make training or behaviour modification quicker and more relaxed.

For horses who are experiencing difficulties such as fear or reactivity, understanding the concept of thresholds and doing your best to keep the horse under threshold while you work through challenges will make the difference between success or no progress.

This traffic light graphic provides another illustration of thresholds.

Fear/Anxiety Threshold

Under Threshold
The horse shows no sign of fear or anxiety towards the trigger.

At Threshold
The horse may move from showing no fear or anxiety to showing some awareness of the stimulus. Lower end of this range: horse can respond to cues and can take treats. Higher end: responses to cues and treats decrease.

Over Threshold
The horse shows signs of fear and anxiety towards the trigger.

Appendix 3
Being a bringer of good things for your horse

Key takeaway points

- *Horses have key biological needs that must be met to ensure wellbeing.*
- *The horse as a complex organism meets these needs by performing specific instinctive behaviours.*
- *We can ensure we have a happy, healthy and willing horse by creating an environment for them in which these needs are met and in which we are A Bringer of Good Things.*

Crucially, our horses will find it easier to meet our "everyday asks" without becoming dangerously frustrated or even aggressive.

What is a horse for?

The horse evolved to be a horse, to survive as a horse by having its horse needs met and to produce offspring horses. That may sound at once somewhat obvious and somewhat philosophical, but it's to make the point that horses are not here to get us round badminton, to win an Olympic dressage medal or even to live in an enclosed field in a non-family group. Therefore, when we as humans create situations in which we aim for such glories, we need to be aware of the tensions that can follow and to do our best to give our equine companions a life that meets their needs in return for what we ask them to do for us.

What is a horse's most basic need?

The fundamental species-specific need of a prey animal such as a horse is to live in a settled, social group where they can take it in turns to be alert for threats and have the ability to run from danger. In evolutionary terms it is essential that this bottom line safety need is met in order that individuals can survive and, crucially, reproduce.

What do "wild" horses do all day?

Horses, as do all animals, have a biologically "pre-programmed" (we might say) activity budget and ethogram (the natural behaviours of a species) which includes:

- grazing/browsing on grass/bushes/trees for up to 16 hours a day
- mutual grooming and play with conspecifics (other horses). Note that actual touch is essential here, not merely proximity.

- moving (walk trot, canter, gallop, buck) as an individual for essential physical exercise, and moving as a group from place to place within their home range
- exploring a variety of objects in a rich environment
- maintenance behaviours, for example rolling, scratching
- loafing/dozing/sleeping (standing/lying on chest/lying flat out) at times and places of their choosing
- understanding "who's who" in the group: who has first refusal of the best sunbathing spot, who gets the best patch of herbage and so on (these will not necessarily be the same horse, and may even change between horses from day to day)
- knowing how each member of the group communicates

What value do humans add?

In traditional horse keeping there are certainly times when we will add good things to our horse's life.

Examples:

- Give tasty hard feed and hay/haylage, for example in winter time (horses living out) when the nutritional benefit of forage is minimal or after several hours without food (stabled horses).
- Give treats at specific times, for example when taking headcollar off at turnout or when our horse has walked up a lorry ramp.
- Randomly give carrots/polos/packaged treats at other unpredictable times.
- Give scratches and massages.
- Provide company on the yard and when exploring out and about on trail rides.
- Provide mental stimulation when training to perform dressage or jumping activities.

In comparison, other aspects of a horse's life often consist of having good things taken away, or not so pleasant things added.

Examples of unpleasant experiences	Why these are unpleasant for a horse
Take them away from friends and forage when we bring them into an isolated stable at night, or for even longer periods of time.	They feel unsafe, scared, anxious. They are forced to stay in an impoverished environment. They become frustrated. They have no choice because they cannot carry out their normal behaviours. Size for size, keeping a horse in a stable is like a human living in a broom cupboard!
Withhold food (this is how the horse perceives the regular routine of evening stables – they are brought in and then have to wait until the owner completes whatever other tasks they deem necessary before a bucket of food appears).	They feel frustrated. They have no choice because they cannot decide when they eat or when to do something else.
Tie them up so they cannot escape, and do things to them that they have not been allowed to properly habituate to e.g. groom, pick up feet, tack up.	They experience fear, frustration. They have no choice because they cannot move away, and are generally punished (shouted at/ smacked) if they move around.

Examples of unpleasant experiences	Why these are unpleasant for a horse
Ride them by applying uncomfortable and painful stimuli e.g. bit pressure, leg pressure, whips, spurs.	They experience discomfort and pain. They have no choice because they cannot escape/avoid the stimuli (to a greater or lesser degree depending on the skill of the rider).
Ride them away from their group and home range e.g. when out hacking.	They feel anxious and unsafe. They have no choice because the rider applies aversive stimuli which they can only escape by complying with the rider's directions.
Take them to shows/endurance rides/rallies.	They feel even more unsafe and anxious because they are outside their home range, even outside the range of sight, sound or smell. Negative reinforcement/positive punishment gives them no choice to avoid this.
Travel them in trailers/lorries.	They are naturally afraid of entering a dark, enclosed space: their eyes cannot adjust to the quick change of light to dark, they are isolated and they cannot flee from threats. After travelling, a horse's sleep is disrupted for up to three days. When horses are travelled regularly they can suffer sleep deprivation. They have no choice but to load because if they don't they will experience pain and discomfort via the application of negative reinforcement/ positive punishment.
Change turnout groups or yards so they are no longer with their friends.	They feel anxious and unsafe because they have left their settled group. They have no choice because humans make these decisions.
Sell them to a new owner – a complete change of environment, removal of their mental map and herd, different training methods applied by a stranger.	They feel even more anxious and unsafe because they have left their settled group, their home range and all the predictable signs that they have learned to pay attention to make sense of their situation. They have no choice because humans make these decisions.

Looking at this as a balance of good and bad, the weighting doesn't look good when we manage horses in the traditional way involving long hours/days in solitary isolation in a boring stable, and when we use riding techniques that utilise uncomfortable training methods. We often prevent horses from carrying out many or most natural behaviours, sometimes we add unpleasant stimuli to a horse's life and we control all the resources that meet a horse's essential needs.

Very often the balance becomes worse in winter, when horses may be kept in all day or longer on occasion, to avoid muddy conditions, for example. Now they have even fewer opportunities for essential interaction with their herd mates, and they spend a longer time in the unenriched environment of their four stable walls – no movement, no grazing, no full contact with other horses, no opportunities to make choices. Added to that, they may be ridden/lunged/ free schooled more in winter to make up for lack of paddock exercise, which might increase unpleasant experiences depending on the type of training methods used.

What effect does this have on our horses?

The horse as a collection of biological systems attempts to cope with a less than ideal environment. However, the unpleasant experiences listed earlier are not simply "unpleasant" as we might understand having a hole in our shoe or not wearing enough clothes on a cold windy day; they are negative experiences because they challenge the horse's ability to cope, both physiologically and mentally. Many horses cope to a greater or lesser degree, but it is always at a biological cost to their physical and mental health:

- Long-term stress:
  This puts stress on all the systems of the body and manifests as both behavioural problems and physical illness including stomach ulcers.
- Learned helplessness:
  Long-term behavioural effects include learned helplessness, where the horse learns that they cannot completely escape aversive stimuli and simply shuts down, doing the bare minimum needed to avoid the worst effects. These shut-down horses are often the horses who can appear to be the most well-behaved!
- Frustration:
  This is seen in a horse that fidgets and won't stand still, or develops stereotypic behaviour such as weaving, wind sucking or crib biting.
- Aggression:
  Some horses don't shut down and instead begin to show aggressive behaviours including threatening and actual kicking and biting.
- Illness:
  Physical effects include increased inflammation in conditions such as lameness or laminitis, and a reduced immune function showing up as skin conditions such as sweet itch.

How can we make a difference? Choice is the key!

You will see that in the prior table, choice is mentioned in every section. For a functioning animal, choice is absolutely essential. Animals in natural environments who have a mental map of their home range make choices all the time: when/where/what to eat, when to move on, where to travel to, when/where to drink, who to play with, what innate motor pattern games to play, when/where to roll, which tree to scratch against, who to mutually groom. Note that horses in a settled group will have a preferred partner, often referred to as a pair bond, who will serve as a regular grazing/grooming/play partner. Removal of this partner, for example because they have a different owner who moves yards, can lead to depression in one or both of the horses in the partnership.

Think about what choices a stabled horse has, with four bare walls, a hay net and a water bucket.

Think about how you could change the aforementioned experiences into good ones, for example when you bring your horse in at night, have a feed ready and waiting for them when they enter the stable. Or, even better, scatter feed hay and hard feed on the floor so that it mimics the horse's natural grazing behaviour. This is good for the health of their spine too.

When we train horses with positive reinforcement (rewards: food or scratches), true we are creating good associations and reinforcing behaviour (*geek point*: the behaviour is reinforced, not the animal), but we are also introducing valuable opportunities for our horses to exercise choice and thus to fulfil an innate need which is crucial to maintaining good mental health.

What do we mean to our horse?

We all talk a lot about what our horse means to us – best friend, comforter, provider of mane to cry into, opportunity for our self-actualisation be it hacking in the woods, winning a first in a county showing class or upgrading in affiliated dressage. But how often do we stop to ask what we mean to our horse?

Animals learn by association, so if something unpleasant is happening to them and something else is "in the picture" at the same time, they may start to associate the neutral "something else" with the specific unpleasant event. For example if they don't like being shod they will begin to associate the farrier's van with that experience and when the van pulls up on the yard the horse may turn away to the back of the stable.

Similarly, if something good happens a horse will associate other things that were around at the time with that good thing. For example when my father used to come to the yard or to see me at competitions, my horse would always seek him out because he knew Dad always meant polos!

The more intensively we manage our horses; that is, the more we keep them stabled and do activities (schooling, ground work, hacking, competitions, travelling, clipping, shoeing, etc.), the more opportunities they have to create qualitative conceptualisations about us. Are we a good thing or are we a bad thing in their life?

We are usually the common factor in everything that happens to our horses, and consequently we have the potential to becoming a good or bad association, depending on the balance of pleasant or unpleasant things that happen to our horses while we are around.

This in itself can create stressful conflict in our horses, because we become both *appetitive* (we give tasty food and we turn our horse out with their friends) and *aversive* (we put our horse in isolation in a boring stable, we create discomfort if we are unbalanced when we ride) in this respect.

A further stressful factor for horses is that they are *affiliative*; they like and need to create good social relationships with whichever species they interact with. It can be a cause of great frustration to them if we are sometimes associated with good things and sometimes with not so good things.

Also remember the phenomenon of learned helplessness; if a horse finds it just too difficult to deal with these conflicting situations they will shut down and appear manageable and compliant, if somewhat dull. Some horse people believe this is the norm for a horse's behaviour, and indeed it can even be desirable because such horses appear to be the safe, "bombproof" ones.

Yet another possibility is that our horse may even become unsafe to ride (rearing, bucking, bolting) or aggressive due to frustration, fear, or a perceived need to guard valuable resources.

A life worth living

We categorise horses as a domestic animal. What does this mean? Tame? Submissive? Trainable? There is evidence that domestication caused genetic changes in horses; however, unlike dogs who through their domestication process arguably gained as much as humans, for individual horses the deal was of far less benefit: first they were hunted for meat, then captured and bred for the same purpose, then trained by force to be used as draught and later ridden animals. Domestication is a thin veneer when it comes to horses.

Consider the condition of zoo animals. Their care and management is governed by the British and Irish Association of Zoos and Aquariums (BIAZA), who understand how compromised a wild animal's life is when kept in captivity. They set high standards that include environmental and psychological enrichment, giving animals choice throughout the day and night.

When the UK government started to evaluate farm animal welfare in the 1960s, Professor Roger Brambell developed the Five Freedoms to which all farm animals should have access:

Freedom from hunger or thirst by ready access to fresh water and a diet to maintain full health and vigour
Freedom from discomfort by providing an appropriate environment including shelter and a comfortable resting area
Freedom from pain, injury or disease by prevention or rapid diagnosis and treatment
Freedom to express (most) normal behaviour by providing sufficient space, proper facilities and company of the animal's own kind
Freedom from fear and distress by ensuring conditions and treatment which avoid mental suffering

These freedoms are now embedded in the Defra Code of Conduct for the welfare of horses.

How many of these freedoms do we give our individual horses? And to what extent?

Researchers such as Bristol University in their Welfare Assurance Programme have gone further and are looking at what animals need in order to give them "a life worth living".

The bank account of good things

It is helpful to look at our holistic relationship with our horse as a kind of bank account. When we bring our horse in from their herd and leave them alone in a stable overnight, that's a debit. When we use a lot of pressure to get a difficult dressage movement, that's a debit. These are the things that horses find difficult to cope with and if this is all that their life consists of they can become shut down, frustrated or aggressive. As animals learn by association, we will become associated with that sense of frustration, so that eventually our horse might even start to feel frustration as soon as they see us.

On the other hand, when we take them to their field to reunite with their friends, that's a credit (although we have to bear in mind the renewed resource holding contests that happen every time a dispersed group is reunited!). When we create stimulating puzzles for them or offer them training choices using rewards, that's a credit. When we help them to use their bodies in easy, pleasant biomechanical ways, that's a credit. When we go for a fun hack with the group from the same field, that's a credit.

We can counteract the possible negative factors that we might bring to a horse's life when we keep them in captivity by making sure that our bank account of good things always has a good credit balance.

Examples of credits we can provide include:

- more hours turnout
- providing company and keeping the herd unchanged
- "one in, all in", all the herd are in or out rather than the unsettling experience of members coming and going at different times
- consider using a barn housing system, where there are no individual stables; horses live as a group and doors are open all the time allowing full choice to be in or out
- when stabled, ensure neighbours can touch and groom each other
- stable pair bonds next to each other
- enrich the stable environment by scatter feeding from the floor (both hay and hard feed; this allows eating on the move which is a horse's natural preference), hanging fruit and vegetable "mobiles" and adding logs for browsing, attaching small coir-type mats or old grooming brushes to the walls to enable self-scratching

- preference test studies have shown that horses favour straw bedding, then shavings. Rubber matting was not very popular as a bedding surface
- reward-based training, including "life skills" such as farrier and vet visits, desensitise to dental gag, farrier's tripod, vet injections and the like

Welfare and behaviour research over many years and many species shows that an animal experiences better biological health and enhanced mental and emotional wellbeing when their natural needs are met. We refer to species-specific needs as the needs that a species spends time meeting in their natural environment. The way an animal spends this time is called an "activity budget" or "time budget", and the more we can make our horse's domestic life match their natural time budget the happier and healthier our horses will be.

When horses spend time "doing" their activity budgets they are all the time making choices. Choice is one of the best credits we can give our horses. A colleague visited a very orderly livery yard run with military precision and not a wood shaving out of place. The only change the yard owners would accept was to offer the horses two water sources in their stable, an automatic waterer and a water bucket. Just adding this single choice to the horse's lives made an improvement to the behaviour of the difficult horses.

By enriching their environment in these ways we give our horses much more choice and keep our "credit balance" high. Choice is a key factor; it's not enough just to train with positive reinforcement, if in effect there is no real choice for our horse.

Our bank balance will change with the seasons too; in winter we might have to work harder to keep credits up if we are increasing debits such as bringing our horses in for longer periods of time, this is where enriching their stable environment really pays off. And if they can't be stabled next to their pair bond, make sure that they both get regular visiting time every day either across the stable door, in a small turnout paddock or going out in hand together.

Summary of how to be a bringer of good things

- Learn about what a horse's natural needs are.
- Evaluate your horse's situation: how many needs are met?
- Aim to meet more needs; if you can't meet them all, concentrate on the ones where you can make things better.
- Review your "bank balance" regularly.
- Minimise unpleasant experiences for your horse and maximise good ones – it's not just about "being nice"; it means we will have a healthy horse who is a nice person to be around, who greets the world with confidence and with whom we can do all the activities we dreamed of!

Many of my behaviour consultation cases are resolved by making horse care and management changes like these, without any need for further behavioural treatment.

Appendix 4
How to enrich your horse's environment

Benefits

- enables a wider range of your horse's natural behaviours
- reduces stress
- helps horses cope with box rest
- develops good quality brain activity to improve performance
- increases the effectiveness of behaviour modification programmes

- Scatter hay(lage) in piles on floor to copy natural "eating-on-the-move" behaviours.
- Scatter hard feed on hay/haylage so that it takes longer to eat.
- Hide carrots, swedes, apples, turnips in haylage to encourage foraging.
- Hang suitable fruit and vegetables at different levels from trees/posts in paddock, and from beams in stables, to replicate browsing behaviour.
- Put logs and branches in paddock/stables to encourage browsing – check the British Horse Society (BHS), Blue Cross or Horse Trust websites for what to avoid. Oak branches are good but remove the leaves and acorns. Willow is good too.
- Hide small amounts of food in accessible places in stable/paddock to encourage foraging behaviour and reduce boredom.
- Fix textured mats to stable wall at "bum height" to enable self-grooming.
- Turn out for longer periods of time – the ideal is 24/7 with good shelter and safety, but if this is not practical then aim for as near to this as possible (not for box rest horses obviously).
- Open the stable door and use a rubber barrier to hook across the doorway. Put hay on the floor in the doorway so your horse can see out while eating.
- If possible, arrange the partition walls so that horses next to each other can see and touch each other. Try to stable field friends next to each other. If one horse is on box rest, bring their friends to visit them and let them chat over the stable door. If the box resting horse is allowed to walk a short way, then take them to visit their friends in other stables.
- Go for nature rambles with your horse where you just lead them out to explore the environment together, stopping when either of you wants to look at or investigate something. Be like you would with a toddler, feign an interest in the things your horse wants to investigate! This will develop independence and build your relationship. If you can, lay a treasure trail first before you bring your horse out, and then pretend to be as surprised as your horse when they come across a pile of chop, or a few scattered grass nuts or hay cobs, or fruit/veg.
- If it's logistically possible, set up a pen outside the stable with temporary fencing, about the same size as the stable dimensions; leave the stable door open so your horse can choose where to be and when to be in one part or the other. This makes a huge difference in your horse's experience of having some choice and control over their environment which is crucial to their wellbeing.
- Offer choice in other ways too, for example a bucket and an automatic water; don't mix feeds but put piles of different feed in different places, offer choices of fruit and vegetables.

Toy box – encourages natural play behaviours to stimulate positive neural systems, promote wellbeing, reduce stress and inflammation and maintain good immune system

1. Horses naturally mouth unusual objects/shapes, for reasons of play and exploration. Set up a "toy box" and give to your horse (in stable or outside paddock on a long line) as many times as possible each day – scatter small pieces of food in it to encourage natural object play. "Toys" can include safe dog toys, teddies or other soft children's toys, cloths, old brushes, an unused bit, plastic bottles. Always supervise this activity for safety.
2. Place safe toy(s) in stable/pen/paddock to allow free object play by choice (change toys periodically to give different sizes, textures, shapes) and supervise for safety.
3. Some horses are a bit apprehensive of the toy box when they first encounter it. If your horse is one of these, take everything out except one item, and add treats. Sometimes even this is too much for some horses, in which case take all the items out and just put a few (small handful) of treats in, do this several times to accustom your horse to the toy box and then add items one by one over a period of days.

Appendix 5
The target game

You will need: treat bag to put treats in, clicker (if you use one), small cone or other target, low-value treats and a different "end session" treat.

A low-value treat is something your horse wants to have but is not the most exciting treat in the world. Hay cobs, grass nuts or chaff are good examples. Training with treats that are too exciting can create frustration.

Objective: to introduce the concept of reward-based training and to stimulate positive neural pathways.

The target game can also be used in conjunction with behaviour modification programmes such as retraining trailer loading or gradual reintroduction to hacking or schooling.

Important:

- Use start and finish signals for each session.
- Every time you end the session, give an "end of session" treat.
- Click or say your marker word, for example "Yes", "Ok", "Good" etc. before you treat, to mark the behaviour you wanted – you are reinforcing the behaviour of touching the target, but by the time you give the treat your horse may have done other behaviours – you need to make sure they know what it was they did that got them the treat, so click/mark the behaviour of touching the target.
- Hide the target and treat bag as soon as the session is finished – then your horse knows that target + treat bag = the game. This avoids frustration at other times and helps your horse know the rules of the game (touch target = treat).
- If your horse isn't interested in treats, or if they find any treats just too exciting, you can give neck/wither scratches (find your horse's "sweet spot" to do this).

Stage 1

1. Have your horse behind a barrier in a safe place where they do not show any anxiety and can remain calm. This can be behind a stable door, a chain across the door, a fence, a gate or behind a pole across two barrels. Stand on the other side of the barrier. Hold the cone near your horse's nose. Click, or say your marker word – "Yes", "Ok", "Good" etc. and give the treat when they touch the cone with their nose.
2. In the same situation, hold the cone to right and left. Click/marker word and give the treat as mentioned earlier.
3. Hold the cone lower, and to the right and left, until your horse is touching it consistently and calmly on the floor to the right and left. (If you are doing this in the stable you will need to open the door and fix a chain, for example, across the doorway.)
4. Repeat (1) to (3) in another of the situations described at (1); for example, change from over stable door to over field gate. Make sure none of the situations cause your horse anxiety or fear. Repeat (1) to (3) in as many of the earlier mentioned situations as possible.

How to give the treat:

- Don't let your horse mug you for treats. You are not an on-demand treat dispenser. Keep your fist closed over the treat until your horse is polite and calm. Give the treat immediately after this happens. If your horse shows calm polite behaviour you must treat; to do otherwise is not fair and you will get a frustrated horse who may nip or bite.
- Note: if your horse takes a long time to be calm when you have your hand closed over the treat, please contact your behaviour consultant as it is essential that your horse can be calm as you click and treat.
- When you give the treat, your horse's head should be in a natural and neutral position, not up or down, and it should be facing forwards. This discourages your horse from always trying to swing their head towards you. You don't have to wait for this position to happen; it is easy to lure your horse by just holding the treat in the position their head would be in to take the treat and they will move their head there to take it.

Stage 2

After a while start to vary the reinforcement schedule:

Gradually after several sessions start to reduce the number of times you mark and reward, so that you are not rewarding every touch of the cone. Move to reinforcing say one in three touches, then one in four, one in five etc., but mix it up to make it random and unpredictable so your horse doesn't know which time they will be rewarded. *But* if this makes your horse frustrated, go back to mark and reward every time they touch the target.

Practical uses of the target game:

Once your horse is consistently touching the target you can use it while leading them; hold the target in front of them so they walk forward to touch it. This can be used to achieve calm walking, for example going through gateways, into trailers and so on.

You can use the target game to encourage your horse to stand calmly while being shod, clipped or injected, for example. This must only be if your horse can stay calm at this time.

You can also use the target to keep your horse calm if you find yourself in a situation where other horses are over threshold, as can happen on livery yards or at shows.

Note: if your horse cannot stay calm even with the target present, then don't use the target or it will lose its good associations.

Appendix 6
Teaching "Find it!" to your horse

Horses are naturally good at following their noses: scentwork for horses is now being used as enrichment and people are even starting to train their horse to track people! This ability to nose out interesting smells can be channelled into a constructive skill with some simple training.

"Find it!" can be helpful in a variety of situations. A quick game of "Find it!" can help nervous horses to relax in new or unfamiliar situations, such as after you've unloaded and are getting ready for a pleasure ride, in a busy competition environment or on a bustling yard. "Find it!" can be a way to help your horse channel excess energy or calm themselves when invited to do so. And "Find it!" can provide important mental stimulation by encouraging your horse to seek further and further around for food treats, which mimics the seeking behaviour their wild ancestors would have done.

"Find it!" is a simple and fun game to teach your horse. Here's how to do it.

How to Teach "Find it!"

Teach your horse to play "Find it!" by encouraging their natural talent for sniffing things out. A verbal or physical signal of the start of the game can help your horse know when it's time to put their nose to the ground and go searching for something interesting.

To start, give the signal or cue – "find it!" – and immediately scatter some treats on the floor. If your horse doesn't understand what you're asking, help them out by pointing at the treats or crouching down and looking interested in them. If your horse is still struggling, you can start by placing food treats in a bucket or trug, as this is something most horses are familiar with. When your horse finds the food, reward them with praise and repeat the process: give the "find it!" cue and scatter some more treats.

For treats, use hay cobs or grass nuts you can mix with chop/chaff, and add a few sliced apples/carrots/any other fresh food your horse likes, being aware of any individual waistline or metabolic needs.

Once your horse catches on and begins to associate the "find it!" cue with snuffling for treats, start to increase the area over which you spread the food, so that your horse is covering a wider distance to find the same number of treats. Eventually you will be able to scatter treats around a whole yard or some tracks, creating an enrichment activity that adds movement and takes a decent amount of time.

You can use "Find it!" at your horse's mealtimes, turning previously easily hoovered up bucket feeds into a form of enrichment and mental stimulation, also providing some physical exercise.

Make sure other horses are managed so that they can't come and muscle in on your horse's activity. You might need to ask your behaviourist or rehab trainer how to set this up and where to put your horse and the other horses.

"Find it!" can also be a useful way to encourage structured greetings or calm interactions with people or other horses (if the latter, "Find it!" should be done with all the horses involved and carefully organised to avoid resource holding contests; ask your behaviourist if you're not sure about this): scatter treats and let your horse find them as a way of bringing their stress levels down in unfamiliar situations. This also works well if your horse has had a bit of a fright or has been trigger stacked (see the information sheet on thresholds for more on this).

Appendix 7
Case history report horse: Amber – difficult to handle

Background/history

Amber, five-year-old warmblood mare, obtained by owner at three-and-a-half years old, at which point she was already being ridden in an outline expected of the stage of training of a six- or seven-year-old horse. Kept on livery yard, turned out with one older horse during the day, stabled overnight in loose box next to turnout companion, substrate rubber matting with light dusting of shavings. Fed haylage and one small feed per day of proprietary mix, in addition to grazing forage. Owner, a dog trainer, had ridden for 20 years, Amber was her first horse.

Reported behavioural problem

Amber had become difficult to ride and was not responding as expected to ridden cues, and owner felt their relationship was deteriorating.

Presenting signs

Amber walked away from owner when they attempted to bring her in from the field, and friends had to corner her in order to put headcollar on.

During the consultation Amber was seen to be very unwilling to lower her head for the bridle to be put on.

Amber turned to face back of stable when owner approached with saddle and bridle.

When ridden, Amber tensed up and became stiff in her gaits when rider asked her to relax and soften her jaw and engage through the back.

Amber did not engage with owner as expected, and became tense and appeared frozen when owner entered stable at any time.

Differentials

Pain: through use of P+ and R–, Amber had been taught to perform school movements in an outline which she was not ready for in terms of physical or mental development, and repeating these exercises over time caused musculoskeletal pain and discomfort.

Conditioned negative emotional response to owner: relating to owner's presence during pain-eliciting activities and classical association with that.

Conditioned negative emotional response to riding equipment.

Rider lack of skill and experience, having previously ridden older, more experienced and less highly bred horses.

Gastro-intestinal (GI) dysfunction, linked to stress, causing conditioned emotional response (CER) to saddle.

Oral disease or dysfunction causing CER to bridle.

Some species-specific needs not met; for example, separated field companion at night resulted in removal of ability to mutually groom and other pair bond activities; rubber matting discourages sternal and lateral recumbency and may inhibit REM sleep.

Behavioural diagnosis/assessment

Amber had developed a CER to owner and to saddle and bridle because of classically conditioned negative associations which had developed because of the contiguity of these environmental stimuli and her physical pain response.

Owner goals

1. To experience a closer and more affectionate relationship with Amber.
2. To be able to ride Amber in balance and in good form consistent with her age and stage of development.

Initial management goals (pending veterinary investigation)

1. To reduce Amber's experience of negative affect by stopping activities associated with relevant conditioned stimuli.
2. To meet more species-specific/activity budget needs in order to enable a wider range of natural equine behaviours, achieve homeostatic effect, reduce stress and develop good quality brain activity to improve learning.

Treatment programme

Treatment: Medical investigations – referral back to vet for physical investigations into concerns regarding pain, based on aversion to bridle, observation of stiff movement under saddle, history of aversive training and known links between stress and GI dysfunction, typically ulcers.

Management (and also foundation for treatment): Replace rubber matting with straw bed, lower dividing wall between Amber and companion when in stable so they can engage in more affiliative behaviours, add daily enrichment activities including toy box to encourage natural object play, scatter feeding (concentrates), feed haylage from floor and scatter. Stop training and handling activities that Amber finds aversive. Stop use of saddle and bridle. Start exploratory and investigative nature walks on long line in hand.

Treatment: Commence target training as an introduction to the use of positive reinforcement and counterconditioning.

Desensitisation and counterconditioning to owner (ongoing via management changes), and to saddle and bridle.

Follow-up

A full veterinary examination revealed a broken tooth which had been causing significant pain, particularly when the bit was in the horse's mouth. This was treated by removal of the tooth and a period of recovery with pain relief. After three weeks following excision and consistent compliance with management recommendations, owner reported that Amber now approached them in the field and seemed happier to see them when approached in the stable. She now looked softly towards owner with ears forward rather than flattening her ears to her head and turning away to face the furthest corner of the stable as she had previously done.

In discussion with the vet, client requested a check-up by a physiotherapist who developed a six-month physical rehabilitation programme for Amber to rectify physical issues associated with over-training as a young horse, to be carried out by owner with regular visits from physiotherapist.

As a dog trainer who used clicker training, owner took to target training with gusto and is now helping other owners on the livery yard to train their horses using positive reinforcement.

Outcome

Owner carried out DS/CC BMP to saddle and bridle to good effect and has now started to reschool Amber under saddle.

Appendix 8
Case history report horse: Oscar – oral repetitive behaviour

Background

Thoroughbred gelding, nine years old, obtained as a five-year-old from a private seller.

Owner bought Oscar to hunt and to event at Novice level and hoping to upgrade. He is kept on a small private yard and belongs to the yard owner. Housing is a 12' × 12' brick loosebox with a typical wooden stable door, top half always kept open. There are stables to both sides housing other livery horses. The horses cannot reach each other to touch noses externally when their heads are over the half stable door, and the internal walls reach from floor to roof. The stable block forms three sides of a rectangle, three stables across the short side and five down each long side.

The livery horses are turned out in small groups on a rota, either morning or afternoon. Oscar is turned out alone in a small paddock as his owner is concerned that he should not injure himself in the field. Owner rides him every day either hacking out or jump schooling, and for flatwork he is schooled by the yard groom to help improve his dressage scores. He is hunted weekly in the season, and is boxed to events within a 100-mile radius (approx.), around 12 times a year.

Oscar has performed the behaviour described as crib biting since his current owner bought him. At his last dental examination the dentist expressed concern that his upper incisors were wearing abnormally due to his cribbing behaviour, and the owner had heard that a behaviourist may be able to help.

Presenting signs

Oscar would place his upper incisor teeth on the top of the stable door, grasp the top of the door, arch his neck, contract lower neck muscles and suck in air whilst making a gulping noise. He would perform this behaviour repetitively whilst he was stabled.

Differentials

- Genetic predisposition – thoroughbreds have been found to be more likely to engage in crib biting behaviour.
- Early/abrupt weaning – associated with change of diet to higher concentrates and less roughage, and significant separation distress.
- Social isolation/lack of turnout, previously or currently, or other unmet species-specific needs and a need to cope with associated stress.
- Stress relief – crib-biters have been found to have higher basal sympathetic activity and may be predisposed to anxiety which may be internalised, presenting as a quiet, biddable horse. Crib-biting may be a coping mechanism particularly if the roughage:concentrates feed ratio

is not balanced. Horses prevented from cribbing, for example by use of an anti-cribbing collar, were found to have higher heart rate and cortisol levels and showed post-inhibitory rebound when the collar was removed.
- Inappropriate diet/restricted forage.
- Gastrointestinal dysfunction including ulcers.
- Boredom – inability to engage in behaviours consistent with the time/activity budget of the species.

Behavioural diagnosis/assessment

Existing crib biting behaviour while stabled is maintained and exacerbated by stress associated with social isolation, inability to engage in species-specific behaviours of the equid ethogram, which Oscar is motivated to express, and physical and mental demands of ridden activities.

Management goals

1. Address underlying discomfort.
2. Reduce stress.

Treatment programme

It is generally believed that horses who have cribbed for many years are unlikely to stop, but it is possible that managing a horse to reduce the stress in his lifestyle may diminish the behaviour.

Treatment: refer to vet to investigate and treat gastric discomfort, including feeding more roughage and reducing concentrates.

Management: avoid preventive interventions such as a cribbing collar, foul-tasting substances painted on the door surface and surgery. Reduce stress by enabling more species-specific behaviours including longer turnout in company (suitably introduced), environmental enrichment in the stable including scatter feeding at floor level and various ways of adding variety to feedstuffs. Remove internal stable walls to half height to enable communication with neighbours. Make the cribbing surface more comfortable by affixing a strip of rubber. Take Oscar for "nature rambles" around the yard, unused fields and tracks to allow exploratory behaviour.

Treatment: increase positive affect/countercondition by starting positive reinforcement training of easy tasks and games and non-contingent associations with biologically relevant stimuli including food and scratches. Redirect inappropriate oral behaviour by encourage innate species-specific oral object play using an equine toy box at least twice daily.

Follow-up

The owner reported back to me infrequently on average every six weeks; they had followed my programme for introducing a new horse to an established group and Oscar was now turned out for longer. Forage feeding had increased and the groom was facilitating enrichment activities most days, including nature rambles. Feeding had changed from hay nets to spreading hay on the floor and scatter feeding less concentrated feed on top. They had put a rubber strip on the stable door; they had not changed the stable walls. A friend who was a member of an online positive reinforcement online training group came to the yard regularly and was working through some easy training exercises with Oscar. At the same time, Oscar's owner had reduced their hunting activity during the winter due to other work demands and this is likely to have reduced some of Oscar's stress.

After six months the reduction in stress from ridden activities, and the positive action taken to implement many of my recommendations, did appear to have had some effect on the frequency of Oscar crib biting. It had not stopped completely and, according to experience and the literature, this was not to be expected. The owner reported that Oscar appeared generally more relaxed, softer and friendlier. After ten months there had been no further reduction in the behaviour.

Appendix 9
Harvey behaviour consultation

Action plan #1

14/2/21

Dear Jane,

Thank you for engaging me to work with you to help Harvey with his problem behaviour and thanks for all the detailed information you provided in your questionnaire. It's great to see how committed you are to helping Harvey, and now you can be confident that over the next few weeks and months you will learn how to help him in the best way possible. It was also great to see the setup you have at your home, really horse-centric!

Here is Harvey's initial Stage 1 action plan; this gives you the essential underpinning knowledge that you need to have in order to become Harvey's at-home therapist, plus the first changes that you need to put into place.

Date of our Zoom call: 10/2/23
Owner name: xxxxxx
Owner phone: xxxxxxx
Owner email: xxxxx
Horse details: Harvey, 9 year old Welsh x TB gelding, 15 hh, bay
Vet details: xxxxxx Info@xxxxxx Mobile: xxxxxxx
Any medical issues: Proximal suspensory desmitis (PSD) diagnosis, treated with plasma injections and INDIBA (a type of therapeutic radiofrequency treatment for PSD), no box rest.

Reported behaviour problem

General stressy behaviour associated with separation from the herd, trailer, spooky, unpredictable, nipping.

Your goals

To be able to travel Harvey independently and trek at different locations alone and in company.

What you have tried to date, to resolve Harvey's unwanted behaviour

Going back to basics in handling.
Backing him up as a correction.

Working with Julia's rehab plan – core exercises, in hand exercises.
Liberty dynamics based on the programme with an online trainer.

Case history background

Harvey was started at an event yard and was evented, hunted and did working hunter type activities between the age of four and seven. Then bought for a 12-year-old girl who quickly lost interest. Then bought by a 20-year-old jockey as a project during lockdown with a view to selling him on; he did hacking and some XC with him, no flatwork.

Harvey could spook at lots of things, less so now than when you got him (good work!). If he sees a sudden thing he will jump sideways; if it's something further away, he freezes. If you tried to encourage him (in riding situations) to go forward, he would sometimes spin round to go the opposite direction.

You have used a loose ring eggbutt snaffle in the past, and now a sidepull bitless bridle, a bareback pad, and total contact saddle; you would ride with a schooling whip.

Touch acceptance (I always ask about this as it is usually a good measure of anxiety levels): Harvey will accept touch on his body, will usually move or pick up his legs and fidget when touched there. He does not like the sides of his head being touched but seems to like a soft brush down the front of his face. In the stable he turns his head to you sometimes with a nippy gesture when you brush or stroke his mane, but in the arena he accepts this. As we discussed, he may associate grooming in the stable with show or hunting preparation in a former life and be more anxious about this. The same association is likely to be weaker or non-existent in the arena.

If another horse is next door, he is pretty relaxed in the stable. He likes to stand looking out. You have no concerns moving around him. If left too long he will pick up anything like brushes and toss them around. This may be a displacement activity (which we often interpret as boredom); a displacement activity is often performed when there is a degree of anxiety and the horse can't do what they actually like to be doing in that situation. For example in Harvey's case he would probably prefer to be out with his friends. He will stress if the other horse is not next door for more than a few minutes so you avoid that situation.

Harvey is ok if Frodo is taken out of the field. He will stress if Jeff or Amanda (Section As) are moved, which you would rarely be doing. When Frodo is returned, Harvey and he can have a bit of nipping and Harvey head snaking.

You give Harvey some Sedalin for shoeing otherwise he is fidgety and tries to get anything within reach into his mouth (probably displacement/self-soothing again), including the farrier's clothes and equipment, and can be nippy so you usually hold him rather than tie up.

Harvey gets Premier Performance calming powder, active ingredients are L-Tyrosine and L-Arginine.

Behavioural assessment

Harvey experiences emotions of fear and anxiety in the following contexts:

- when separated from other horses
- when you attempt any grooming or tacking up
- being shod

The motivation underpinning his behaviour is that he wants to avoid the situations that cause him to experience fear or anxiety, and he attempts to do this by increasing distance from aversive (unwanted, unpleasant, frightening) situations by nipping, pulling away when being led, fidgeting when restrained or moving away from you in the stable.

He is also motivated to decrease distance from things he wants to be close to; this is why he appears stressed if the ponies leave the field and nips if he is taken out of the field and the ponies are left behind.

Likely contributing factors from information so far

Fight vs flight

Horses can nip in play, and they can also do it as a fear- or anxiety-based behaviour; in fear, they want to make something they perceive as scary or threatening stay at a distance, or go further away, so they perform what are called distance-increasing behaviours in an attempt to achieve this. Physiologically, the sympathetic branch of their autonomic nervous system has become aroused, activating the fight or flight processes. Horses as a prey species would typically default to flight, but often in a domestic situation they don't have that option, for example they are tied up, held on a lead rope or confined in a stable, so then fight and the consequent threatening behaviour, for example nipping, may be expressed. If the trigger (the scary thing) does go further away either by accident or design, they learn that their strategy works; therefore they will repeat it in similar situations.

How do we know that this behaviour is not play, for Harvey? Because of the contexts in which he expresses this behaviour – it seems to be in situations where a horse would typically be stressed or anxious, taking into account the information you have already given me about Harvey.

How Harvey was started and early ridden life

Traditional starting and schooling with lack of understanding of thresholds and how horses learn, therefore likely confusion and build-up of anxiety and stress.

Medical

Discomfort from PSD.

Life changes

Multiple changes of owner and home range add to anxiety and stress.

Sensitisation

Sensitisation to certain situations, for example absence of other horses, loading.

Sensitisation occurs when successful habituation fails to take place. Horses learn to *habituate* to (get used to/become familiar with) various sights, sounds and experiences of life, whether as youngsters or later in life. But if for whatever reason, instead of getting used to something, they become anxious about it or worried, this means they have become *sensitised* in psychological terms. Then, if nothing is done to *desensitise* them, they will become more sensitised, and problematic behaviours can emerge.

Conditioned emotional response (CER)

A CER develops when a horse has frightening, worrying or painful experiences and creates a negative mental association with "the picture" that pertained at the time. Then when similar "pictures" present, the horse experiences the same CER and reacts in a similar way.

Being trigger stacked and going "over threshold"

The attached information sheet on thresholds explains this. Very briefly it is a cumulative stress response to a number of triggers that occur over a period of time. The first few triggers may cause a degree of stress but not enough to cause the horse to go over their coping threshold, but once a few triggers have been "survived" the next one may cause a reaction.

In the case of a conditioned emotional response (see earlier), the triggers will be parts of the "picture" that the horse perceived at the time of the scary event or events. (These are called conditioned stimuli in psychological terms.)

Rehabilitation attempt

Attempts to deal with spookiness using aversive methods recommended by a trainer.

Individual differences

Every horse has a different temperament and different coping levels which affect how they experience their life being stabled, handled, schooled and ridden.

How you will change Harvey's behaviour – the underpinning principles

In order to achieve your goals, our aim needs to be to get Harvey to a mental place where he can be relaxed, and therefore not doing anything problematic because of stress, and he can stay calm enough to do what he is asked, so that you can retrain him and you will be able to reach your goals with Harvey.

I work on a three-stage behaviour change programme, focusing on Resetting a horse emotionally first and then developing their Relaxation and Resilience, so that they are no longer motivated to perform unwanted behaviours, and desired behaviour such as accepting husbandry treatment will be easier to train.

1. *Reset*: this first stage is about building a strong basis of safety and predictability for the horse, and avoids the stress triggers that have caused them to over-react.

   Part of Harvey's behaviour modification programme needs to involve the process of decoupling the connections in his brain whereby he has learned to react aversively to objects or experiences he finds worrying or unpleasant, and an "only good things" approach will start to build his feeling that he is safe and that he can trust you and remain calm. More on this later.

   All horses need to feel safe and secure as an essential underpinning of all the other elements in their life, and this usually needs to be strongly established before moving on in a behaviour programme; Harvey already scores well on this because of the natural way in which he is kept, including 24/7 turnout and a consistent group of friends.
2. *Relaxation*: this stage focuses on developing a horse's ability to relax more often and in a wider range of situations, and adding positive horse-centric activities to their daily life. This stage incorporates some positive exercises and enrichment that will develop stronger coping abilities in different situations, preparing the horse for stage three, Rebuilding Resilience. Again, many of Harvey's needs are already met in this respect, although I have added at least a couple of additional activities in this first action plan.
3. *Rebuilding Resilience*: I move clients on to this third stage when their horse has made good progress in stage two and is showing a regular and frequent relaxed outlook. This stage uses the reliable psychological process of "desensitisation and counterconditioning" (sorry for the long words! See attached information sheet), using the new and stronger levels of relaxation and resilience that the horse has developed. Using this tried and tested method

we aim to change the horse's emotional response from anxious/worried/fearful to calm and relaxed. In the situations where a horse was previously anxious, it will usually then be possible to train them to respond positively to specific cues or situations that will help to manage their health and behaviour.

Recommendations, stage one

The recommendations in this first stage are all aimed at returning Harvey to a better emotional balance and improving his "hedonic budget" (aka getting a better credit balance in his equine bank account!)

Please stop all of the following:

- Carrying a whip when handling or working – he is likely to have a negative emotional response to this so it will add anxiety and our aim is the opposite – to reduce anxiety and have a calm, trainable horse.
- Backing Harvey up as a correction, for the same reason as earlier.
- Use of ah-ah! or shout or a foot stomp. Same reasons as earlier.

Our aim is to get Harvey to a place where he is not expressing unwanted behaviours, and therefore does not need to be corrected. This way, it is then easier to train him to perform the behaviours you do want him to do.

Loading

If the loading retraining is continuing to improve, please continue. If it gets worse please let me know. Please read the handout "Desensitisation and Counterconditioning" which explains the theory behind what you are doing with the loading retraining, and you are doing well with this so far. There is a second handout with this, called "Gradual Exposure Therapy for Unsettled Horses". Just read this for now and ask me any questions. Although it refers to food treats these are not essential and indeed sometimes I will not recommend that they are used in the way described in the handout – the handout just explains the principle. You are already using food (not hand fed, it's in a bucket) in Harvey's loading rehabilitation and this seems to be going ok at the moment so that's good.

Avoid the arena temporarily

Ideally I would like Harvey to have a two week "holiday" from work in the school. If this is ok for you, please give me a shout at the two week point so I can tell you how to gradually reintroduce this.

This also answers your question about starting ridden work. Restarting ground work and then ridden work are the next two stages in this behaviour programme so please avoid both of these for the next two weeks if you can bear to (it will pay dividends) then send me a progress update and if all is well you can restart groundwork; if his stress seems lower in that activity we can plan to restart ridden work soon after.

Enrichment

This is the secret sauce that will start to really help Harvey's relaxation and resilience in general, and this will have a beneficial effect in reducing his stressful behaviour. These activities are ideally done on a daily basis. This is what helps our horses to be more responsive to us and to feel more settled, less likely to react to triggers that would previously have set them off.

Harvey already has a lot of enrichment because of the natural life that he lives. Two additions I recommend are nature walks and a daily toy box, both are explained in the attached information sheet "Environmental enrichment – horses".

If you haven't already joined, this Facebook group has more great enrichment ideas (you can never do too much!) www.facebook.com/enrichmentequine

Touch acceptance programme

Please read the attached training guide "Touch acceptance" and begin to work through it. Harvey is unsettled about being touched in some areas, and working through a touch acceptance programme is beneficial for all horses for the reasons explained in the handout. However, if he appears to regress stress-wise, please stop this temporarily and we will come back to it later.

Next steps

We have arranged our follow-up call for three weeks' time, and please do check in with me at any time via WhatsApp.

Moving on

Activities and behaviour programmes we will move on to:

- Groundwork
- Ridden work

Desensitisation and counterconditioning to:

- Leaving the herd (i.e. becoming more comfortable with this).
- Tack (if necessary – we'll need to test this when you move on to the next stage of restarting groundwork, depending on what additional tack you use other than a headcollar, and the same tests for saddle and bridle when you start riding him again).

Reading material and activities

Please start to implement the enrichment activities (nature walks and the toy box), from the Environmental Enrichment handout.

Start to implement the Touch Acceptance programme (see attached activity sheet) but if Harvey appears to regress stress-wise, please stop this temporarily and we will come back to it later.

Please read all the attached information and activity sheets:

How horses learn
The equine bank account
Thresholds
Environmental Enrichment
Touch acceptance
Desensitisation and counterconditioning
Gradual exposure therapy for horses
Start and stop signals for training or rehab sessions
The Find It! game (another enrichment piece)
Evolution
The social structure of horses
Key equine needs

I also recommend a very good equine behaviour book called *Equine Behaviour in Mind: Applying Behavioural Science to the Way We Keep, Work and Care for Horses*.

Any questions?

Please let me know if you have any questions at all.

Best wishes,

Appendix 10
Sam behaviour consultation

Action plan #1

14/4/24

Here is Sam's initial Stage 1 action plan; this gives you the essential underpinning knowledge that you need to have in order to become Sam's on-the-yard therapist, plus the first changes that you need to put into place.

How you will change Sam's behaviour – the underpinning principles

Our aim needs to be to get Sam to a mental place where he can be relaxed, and therefore not doing anything problematic because of stress, and he can stay calm enough to do what he is asked, so that you can retrain him and you will be able to reach your goals with Sam.

I work on a three-stage behaviour change programme, focusing on Resetting a horse emotionally first and then developing their Relaxation and Resilience, so that they are no longer motivated to perform unwanted behaviours, and desired behaviour such as accepting husbandry treatment will be easier to train.

1. Reset

This first stage is about building a strong basis of safety and predictability for the horse, and avoids the stress triggers that have caused them to over-react.

Part of Sam's behaviour modification programme needs to involve the process of decoupling the connections in his brain whereby he has learned to react aversively to objects or experiences he finds worrying or unpleasant, and an "only good things" approach will start to build his feeling that he is safe and that he can trust you and remain calm. More on this later.

All horses need to feel safe and secure as an essential underpinning of all the other elements in their life, and this usually needs to be strongly established before moving on in a behaviour programme; Sam already scores well on this because of the natural way in which he is kept, including 24/7 turnout and a consistent group of friends.

2. Relaxation

This stage focuses on developing a horse's ability to relax more often and in a wider range of situations, and adding positive horse-centric activities to their daily life. This stage incorporates some positive exercises and enrichment that will develop stronger coping abilities in different situations, preparing the horse for stage three, Rebuilding Resilience. Again, many of Sam's needs are already met in this respect, although I have added at least a couple of additional activities in this first action plan.

3. Rebuilding Resilience

I move clients on to this third stage when their horse has made good progress in stage two and is showing a regular and frequent relaxed outlook. This stage uses the reliable psychological process of "desensitisation and counterconditioning", using the new and stronger levels of relaxation and resilience that the horse has developed. Using this tried and tested method we aim to change the horse's emotional response from anxious/worried/fearful to calm and relaxed. In the situations where a horse was previously anxious, it will usually then be possible to train them to respond positively to specific cues or situations that will help to manage their health and behaviour.

Recommendations, stage one

The recommendations in this first stage are all aimed at returning Sam to a better emotional balance and improving his "hedonic budget" (aka getting a better credit balance in his equine bank account!).

Please stop all of the following:

- Carrying a whip when handling or working – he is likely to have a negative emotional response to this so it will add anxiety and our aim is the opposite – to reduce anxiety and have a calm, trainable horse.
- Backing Sam up as a correction, for the same reason as earlier.
- Use of "ah!", shout or foot stomp. Same reasons as earlier.

Our aim is to get Sam to a place where he is not expressing unwanted behaviours, and therefore does not need to be corrected. This way, it is then easier to train him to perform the behaviours you do want him to do.

Avoid the arena temporarily

Ideally I would like Sam to have a four-week "holiday" from work in the school. If this is ok for you, please give me a shout at the three-week point so I can tell you how to gradually reintroduce this.

This also answers your question about starting ridden work. Restarting ground work and then ridden work are the next two stages in this behaviour programme so please avoid both of these for the next four weeks if you can bear to (it will pay dividends), and then send me a progress update and if all is well you can restart groundwork; if his stress seems lower in that activity we can plan to restart ridden work soon after.

Enrichment

This is the secret sauce that will start to really help Sam's relaxation and resilience in general, and this will have a beneficial effect in reducing his stressful behaviour. These activities are ideally done on a daily basis. This is what helps our horses to be more responsive to us and to feel more settled, less likely to react to triggers that would previously have set them off.

Sam already has a lot of enrichment because of the natural life that he lives. Two additions I recommend are nature walks and a daily toy box, both are explained in the attached information sheet "Environmental enrichment – horses".

Touch acceptance programme

Please read the attached training guide "Touch acceptance" and begin to work through it. Sam is unsettled about being touched in some areas, and working through a touch acceptance programme is beneficial for all horses for the reasons explained in the handout. However, if he appears to regress stress-wise, please stop this temporarily and we will come back to it later.

Moving on

Activities and behaviour programmes we will move on to:

- Ground work
- Ridden work

Desensitisation and counterconditioning to:

- Leaving the herd (i.e. becoming more comfortable with this).
- Tack (if necessary – we'll need to test this when you move on to the next stage of restarting groundwork, depending on what additional tack you use other than a headcollar, and the same tests for saddle and bridle when you start riding him again).

Appendix 11
Shaping plan – desensitise my horse to being saddled

GOAL BEHAVIOUR (what is the goal behaviour you want to achieve?):	My horse stands calmly while I saddle him up.
WHY do you want to shape this behaviour?	So that my horse feels calm and relaxed about being saddled, because I plan to ride him once this is achieved.
WHERE will you do this? Do you need to use a protective barrier? And, where will you generalise this behaviour in later stages?	1. In the stable 2. Tied up outside the stable 3. Leading round the yard 4. Leading up and down lane
WHO will be involved? (Do you need an assistant?)	Me (one handler)
WHAT equipment do you need?	1. Handkerchief (stage 1) 2. Tea towel (stage 2) 3. Numnah (stage 3) 4. Saddle (stage 4) 5. Bag to hold food treats
WHAT treats or other rewards will you use?	Grass nuts and chaff
WHEN will you do this?	Before morning turnout and after evening stables
HOW will you reinforce the correct behaviour?	Say "yes!" and give a handful of chaff/grass nuts, enough to elicit good chewing

Step number	**What you want your horse to do** **(OR you can write this as what you will do)** ***GIVE A FOOD REINFORCER AFTER EACH REPETITION***	**Tick (or put Y) when achieved, and move on to the next step**
1	Put handkerchief on his back and he stays calm, from both sides.	
2	Put tea towel on his back and he stays calm, from both sides.	
3	Target to numnah (see target training shaping plan, using numnah as target).	
4	Target to saddle (see target training shaping plan, using saddle as target).	
5	Put numnah on his back and he stays calm, from both sides	
6	Put saddle on his back and he stays calm, from both sides	
7	Repeat steps 5 and 6 while tied up outside stable.	
8	Repeat steps 5 and 6 while leading round yard.	
9	Repeat steps 5 and 6 while leading down lane.	
10		
11		

Index

Note: Page numbers in *italics* indicate a figure and page numbers in **bold** indicate a table on the corresponding page.

For Product Safety Concerns and Information please contact our
EU representative GPSR@taylorandfrancis.com Taylor & Francis
Verlag GmbH, Kaufingerstraße 24, 80331 München, Germany